Precalculus Investigations

A LABORATORY MANUAL

PRELIMINARY EDITION

Gary M. Simundza • Charlene R. Solomon • Anita A. Penta
Robert C. Cournoyer • Dwight F. Horan
Richard T. Goller • Marcia J. Kemen

Wentworth Institute of Technology

PRENTICE HALL
Upper Saddle River, NJ 07458

Acquisitions Editor: Sally Denlow
Production Editor: Robert C. Walters
Special Projects Manager: Barbara A. Murray
Cover Designer: Joseph Sengotta
Manufacturing Manager: Trudy Pisciotti
Cover Photo: Circular staircase in the Old State House, Boston, Massachusetts
by Barbara Simundza Studios, courtesy of The Bostonian Society/Old State House

Simon & Schuster / A Viacom Company
Upper Saddle River, New Jersey 07458

Printed in the United States of America

10 9 8 7 6 5 4 3 2

ISBN: 0-13-010954-1

PRENTICE-HALL INTERNATIONAL (UK) LIMITED, LONDON
PRENTICE-HALL OF AUSTRALIA PTY. LIMITED, SYDNEY
PRENTICE-HALL CANADA INC. TORONTO
PRENTICE-HALL HISPANOAMERICANA, S.A., MEXICO
PRENTICE-HALL OF INDIA PRIVATE LIMITED, NEW DELHI
PRENTICE-HALL OF JAPAN, INC., TOKYO
SIMON & SCHUSTER ASIA PTE. LTD., SINGAPORE
EDITORA PRENTICE-HALL DO BRASIL, LTDA., RIO DE JANEIRO

Table of Contents

Preface to the Instructor

These Mathematics Laboratory Investigations are intended to be an integral part of a two-semester algebra/trigonometry/precalculus course sequence for college freshmen of average mathematical ability. The authors have seen our own students' enthusiasm for mathematics considerably heightened because of their response to these authentic problems from a wide variety of professional workplace situations.

The activities were designed for use in an extended class period, typically a two-hour laboratory component of a standard mathematics course, although many can be done in two consecutive one-hour classes. Several involve laboratory experimentation that may be best accomplished in a physics or electronics lab, while others require only taking measurements from supplied charts and can be done in standard classrooms. It is expected that students will work on labs in groups, and will use graphing calculators or computer software as part of their analysis.

Each laboratory investigation is based on a well-defined goal, whether it be to design a staircase to fit an interior space while satisfying building codes, to create a set of instructions for cutting out a wrench shape on a milling machine, or to make an optimal choice between alternative office rental spaces or equipment purchases. Particular mathematics objectives are addressed in the context of achieving these goals, while students are led to model problems analytically, numerically, and graphically as well as being challenged to articulate their understanding of mathematics concepts in writing.

Brief Instructors' Notes are provided for each laboratory investigation, with tips on use of equipment and/or pedagogical hints based on our own classroom experiences. Many labs have "Extensions" listed at the end, which suggest further activities that students can engage in for enrichment purposes. We often assign these as optional "extra credit" tasks for motivated students.

Instructors will find it essential to complete the labs themselves in advance in order to be able to anticipate student questions and to help students with equipment setup and data collection. The authors will be happy to assist with any questions of your own that may arise with respect to any of the labs, and can be reached by e-mail or by telephone (see the Instructors' Table of Contents for the name of the author of each laboratory investigation):

Bob Cournoyer	cournoyerr@wit.edu	617-989-4356
Dick Goller	gollerr@wit.edu	617-989-4361
Dwight Horan	horand@wit.edu	617-989-4358
Marty Kemen	kemenm@wit.edu	617-989-4352
Anita Penta	pentaa@wit.edu	617-989-4351
Gary Simundza	simundza@wit.edu	617-989-4354
Charlene Solomon	solomone@wit.edu	617-989-4355

Teaching mathematics with Mathematics Laboratory Investigations, as with any innovative approach, demands work on the part of the instructor. But we have found the rewards, both for students and instructors, to be well worth the extra effort involved. We trust that you will, too, and welcome you to our group of instructors of mathematics as a laboratory science.

Contributors

The following people worked hand in hand with us on the initial drafts of our laboratories. We greatly appreciate all of their contributions to the contents of these laboratory investigations.

Kurt G. Benedict, Registered Professional Engineer
Department of Architecture

Kenneth E. Bourque
Department of Electronics and Computer Science

Eleanor K. Canter
Department of Applied Mathematics and Sciences

Frederick E. Gould, Registered Professional Engineer
Department of Civil, Construction, and Environment

Francis J. Hopcroft, Registered Professional Engineer
Department of Civil, Construction, and Environment

Suzanne L. Kennedy, Certified Facilities Manager
Department of Design and Facilities

Thomas M. Lesko, Registered Architect
Department of Architecture

Kathryn Nimeskern O'Neill, Certified Manufacturing Engineer
Department of Mechanical and Manufacturing

Peter S. Rourke, Certified Manufacturing Engineer
Department of Mechanical and Manufacturing

James M. Winter
Department of Mechanical and Manufacturing

Acknowledgements

The authors are grateful to the administration of Wentworth Institute of Technology for their support of this interdisciplinary effort. Its provost, George T. Balich, and other academic leaders of the Institute have promoted the diversity of expertise and talents brought by individual faculty to the project team, to which the project owes its success. By facilitating pilot-testing of the laboratory investigations in Wentworth mathematics classes, Provost Balich and Department Head Wilfred Caissie allowed team members to evaluate their utility and effectiveness, as well as to accumulate invaluable student feedback and suggestions.

In addition to the Contributors whose names appear inside the cover, and who were involved directly in the initial creation of these laboratory investigations, the authors wish to thank the following individuals for sharing their expertise in early discussions of ideas for lab activities: Frederick Driscoll, John Marchand, and Robert Villanucci of Wentworth's Electronics and Computer Science Department; Herbert Fremin and Rachel Pike and of the Design and Facilities Department; Thomas Taddeo of the Civil, Construction, and Environment Department; Charles Cimino, Phillip Comeau, Steve Diamond, Garrick Goldenberg, Jeffrey Stein, and Dr. Glenn Wiggins of the Architecture Department, retired FAA sector chief Phillip Nimeskern; Dr. Glenn Pavlicek, Professor of Mathematics at Bridgewater State College; and David Leo, senior mechanical engineer at General Electric Company. Also Carol Rogers of Vermont Technical College, Robert Clark of Adirondack Community College and Douglas Holley of Hingham High School for their insightful reviews of early drafts of many of the labs. And many thanks to our Advisory Board for help in guiding the direction of the project: Mary Canty of the Boston Public Schools, Jane Devoe of Northeastern University, Dr. David Entin of New York City Technical College, Alan Hadad of the University of Hartford, Dr. Fadia Harik of Bolt, Beranek, and Newman Company, Robert Kimball of Wake Technical Community College, Lynn Mack of Piedmont Technical College, John McDonagh of the Massachusetts Center for Career and Technical Education, Jean Simcic of Schenley High School, and Dr. John Tobey of North Shore Community College. We appreciate the assistance we received from many individuals on the Wentworth staff, particularly in the Communications Center, the Computer Center, the Alumni Library, the electronic labs (particularly Bob Carlson), and our administrative secretary Regina Stals Ramos. Finally, we'd like to thank our students for their enthusiasm, and for helping to show us what worked and what didn't.

Photography for "Design of Spiral and Circular Stairs" and "Design of a Straight Staircase" was provided by Barbara Simundza Studios.

The development of these laboratory investigations was supported by a generous grant from the Advanced Technological Education program of the National Science Foundation.

Preface to the Student

The Mathematics Laboratory Investigations in this book are based on real workplace problems encountered by people in a variety of occupations. Some problems have been somewhat simplified or idealized in order to make the mathematical connections clearer, but the fundamental nature of each application has been preserved. In working through these labs you will see how thinking mathematically can help practicing professionals deal efficiently and effectively with problematical situations that arise in the course of their work. You'll be expected to work on each problem in a group with other students, similar to the way people typically collaborate in teams in their jobs.

Suggestions for Success

- You'll probably find that thorough reading of the introductory material in each lab, as well as careful attention to the wording of the questions that are posed throughout, are keys to understanding and successfully completing the lab.
- In labs that involve experimentation, ask for assistance if you have any uncertainty about equipment setup and data collecting.
- Most labs assume that you have a graphing calculator available for your use.
- You'll be expected to be able to rely on mathematics skills you have previously learned. These are usually specified under "Prerequisites" at the beginning of each lab.
- Some of these laboratory activities will introduce you to new mathematics topics. Don't be surprised if you encounter an unfamiliar concept; the lab will help you learn and understand it.
- Many questions require written responses, in which you are asked to demonstrate your understanding of mathematical concepts. Discuss these among your group, and write in clear and complete sentences.

We hope that you will find these laboratory investigations challenging—they are designed to make you think, sometimes creatively, about mathematics—but that you will also get a great deal of satisfaction from using your mathematical knowledge to tackle these problems.

MATHEMATICS LABORATORY INVESTIGATION

AIRCRAFT NAVIGATION I

Topics: **CHART READING, INTRODUCTION TO VECTORS**
Prerequisites: *Cartesian Coordinate System*
Equipment: VFR paper and laminated charts, compass or dividers, dry erase markers

I. INTRODUCTION:

A student pilot must study charts and navigation in order to get a Private Pilot's License. Included in the requirements are both a written and a flight exam. These exams include chart reading. Primarily two charts are used - a sectional aeronautical chart used for *Visual Flight Rules* (VFR) and an enroute chart used for *Instrument Flight Rules* (IFR). IFR is used when weather and visibility are poor, during times of high air traffic, and for all commercial flights.

The region to become familiar with today is the Boston area airport, called Logan Airport. Included here for study is a VFR terminal chart of Logan Airport and a Legend of the symbols used on the chart.

II. USING THE VFR (Visual Flight Rules) TERMINAL CHART:

Let's start by looking at the VFR Terminal Chart. The pilot uses geographic references to fly so flight information is overlaid on a topical projection of the land. Look over the VFR chart and locate Logan Airport. Notice that it is centered inside several overlaid concentric circles surrounded by an overlaid rectangle. The region enclosed by the rectangle is considered to be the *terminal area*. Now look left and find Fort Devens. It lies along one of the concentric circles.

The legend on a chart defines the symbols used in the chart. On last page of this lab where you will find a portion of the VFR legend. Look over the legend at the variety of symbols used. **Now pick a symbol and find two of those objects. Circle** these objects on your VFR chart.

What symbol and objects did you choose?

Describe their general location on the chart.

LATITUDE AND LONGITUDE

It is more accurate to describe the location when a systematic grid is overlaid on the chart. As you were looking at the chart did you notice that it is separated into rectangular grids of latitude and longitude by lines that are evenly tick marked? When looking at the entire chart, the divisions for these grids are best observed along the edges of the chart.

To make them easier to see and use, **find** the 42° and 43° (horizontal) latitude lines and **mark** them along the left and right edges of your VFR chart. Then **find** the 71° and 72° (vertical) longitude lines and **mark** them at the top and bottom of your chart. Each primary marking represents one degree while the secondary markings represent 30 minutes, or half a degree.

What unit does the smallest division on this chart represent?

Logan Airport is located at a latitude of approximately N 42^0 21.45' and a longitude of approximately W 70^0 59.37'. **Check it out.**

Describe in the table that follows, the location of the objects chosen above using latitude and longitude.

Name of Object	Latitude	Longitude

Land and Nautical miles

A *nautical mile* is used for both air and sea navigation because it is related to the latitude and longitude lines on the chart. At the equator each degree of longitude represents 60 nautical miles. Everywhere each degree of latitude represents 60 nautical miles. A nautical mile (nm) equals 1.15 land, or statute, miles.

How many nautical miles are represented by each minute of latitude?

Notice that in a terminal area around Logan there are circles of varying radii that can help you determine how many nautical miles a landmark is from Logan. Once again **look to the West** and now **determine the number of nautical miles** from Fort Devens to Logan Airport.

Calculate the number of land miles between Fort Devens and Logan Airport.

Among the information included on the VFR Terminal Chart are airport runway configurations, elevation information, and key IFR (Instrument Flight Rules) airways.

ELEVATION

Inside each latitude/longitude grid sections is a number representing the maximum elevation of the tallest object in the grid. The full size numbers represent thousands of feet while the half size numbers represent hundreds of feet. For example, the grid section bounded on the bottom by a latitude of N42° 30', on top by 43°, and on the right by a longitude of W72° has the number **3**^{5} within it, telling a pilot that the highest object is at an elevation of 3500 feet.

What would you use as the minimum flying altitude to safely clear all objects in the entire terminal area (enclosed by three sides of a rectangle)**?**

At what elevation is the tower near Hudson (which is southeast of Fort Devens)?

What is the height of the tower? (Hint: the legend should tell you how to distinguish between numbers by the tower.)

Would it have been cleared flying at the minimum altitude above**?**

Look at the runway configuration of Logan Airport and **compare** it to the airport diagram on the right.

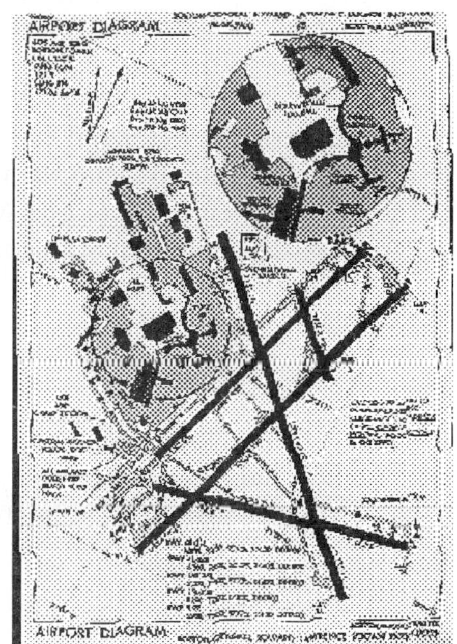

Now **find** the airport at N 42^{0} 33.5' and W 71^{0} 46'.

What is its name?

Find Shirley airport. Notice how close together the airports are.

How could a pilot tell from the air which airport is which?

Sketch the configuration of the runways.

III. KEY IFR AIRWAYS - VECTOR ROUTES (ON THE VFR CHART)

Although this is a Visual Flight Rules chart, the key IFR (Instrument Flight Rules) airways appear on the chart. They are drawn as shaded lines and can be most easily seen by looking at a *compass rose* around an airport. The arrow within the compass rose indicates the direction of *magnetic north.*

An IFR airway on a Terminal Chart is identified by its initial direction and number. All low altitude airways begin with the symbol V. Because these airways have both length (magnitude) and direction, they are *vectors*. The vector number is called the name of the route. On this chart the direction of the vector is noted in degrees along the vector line close to the airport configuration. The length may be enclosed in a box above or below the vector name. Because of the direction of the prevailing winds around Boston, V-141 is a key route for Logan Airport..

Find the initial direction of V-141.

Assuming that the directions refer to outgoing airway routes, **trace along** the other vector airways around Logan Airport**.**

Name another vector route leaving Logan.

What is its initial direction?

Suppose you were flying in an aircraft along V-141 and were able to see directly behind you, **in what direction would you be looking**?

Planes can be vectored in or out along the same airway (however, not at the same time).

History has proven that it is easy to get turned around when flying. Have you heard of "Wrong Way Corrigan" who wanted to fly to Dublin, Ireland but was denied permission to make such a dangerous flight? On July 17, 1938 he filed a flight plan to fly from New York to Los Angeles but landed instead in Dublin Ireland claiming that he had made a common calculation error. What error did he claim?

Comparing vectors.

Now look at the compass rose around Gardner Airport and locate magnetic north (0° indicated by the arrow). Read around the compass rose in a clockwise direction. Notice that 90^{0} is indicated by a 9 and 180^{0} by an 18.

In the table that follows, list all the Vector Airway Names and the initial direction for each airway.

Airway Name	Direction
V-39	42°

Now **draw a vector** from the airport configuration at Gardner to the airport configuration at Bedford. Start drawing a ray at the initial point and place an arrow at the endpoint (——→), indicating the direction of the vector.. A vector not only has direction, but also has *magnitude.* Using the latitude grid, **approximate its magnitude** and using the compass rose, **approximate its direction.**

Next examine the distance from Gardner to Bedford along V-431. To measure the length of the vector, place one end of your compass on the VOR symbol at Gardner (shown above) and the other end at the center of the airport configuration at Bedford. The length of the opening between points of your compass is the magnitude of the vector and represents the number of nautical miles between Gardner and Bedford.

Two vectors are equal if they have the same direction and magnitude.

On your VFR chart**, draw a vector** beginning at Logan Airport that is **equal to** the vector from Gardner to Bedford. If there is a landmark near where the vector ends, name it. If not, give its position by stating its latitude and longitude.

Now on the chart, **draw a vector** beginning at the VOR outside Keene **equal to** the vector from Gardner to Bedford. If there is a landmark near where the vector ends, name it. If not, give its position by stating its latitude and longitude.

Consider the vector along V-431 from Gardner to Bedford and the vector along V-431 from Bedford to Gardner. **Are these vectors equal? Why or why not**?

Consider the vector along V-431 from Gardner half way to Bedford.

Where does it end?

Is it equal to the vector from Garner to Bedford? **Why or why not?**

Write a question for your group to answer that illustrates the concept of equal vectors.

Write a question for your group to answer that illustrates the concept of unequal vectors.

IV. SUMMING UP

Vectors are used in many application such as Seacraft Navigation, Robotics, Structural Engineering, Voltage-Current relationships in Electronics, and Surveying. Like Aircraft Navigation, Seacraft Navigation states the direction of the vector using a compass rose, but many other applications do not.

List some places where you have seen or used vectors in the past.

Were the directions of the vectors given using a compass rose?

To prepare for some of the applications that state the direction of a vector without the compass rose, use the rectangular coordinate system below to draw the following vectors.

Before you graph the vectors on the axes provided below, add in both a horizontal and vertical scale. For each vector, the endpoint with an arrow (⟵) indicating its direction. Label each vector.

Draw a vector, V_1, 90° from north that is 5 nm long. Placing its initial point at the origin and **identify** the coordinates of its endpoints.

Next place its initial point at the origin and **draw** V_2, which is 180° from north and 4 nm long. **Identify** the coordinates of its endpoint.

Now **draw** V_3, beginning at (-3, 1) and **equal to** V_2

Draw V_4, beginning at the origin and 180° West of East and 6 nm long.

Finally **draw** V_5, beginning at the origin 45° North of East about 1.5 nm long.

N

SOUTH | NEW YORK | NORTH

NEW YORK
LEGEND

Airports having Control Towers are shown in Blue, all others in Magenta. Consult Airport/Facility Directory (A/FD) for details involving airport lighting, navigation aids, and services. For additional symbol information refer to the Chart User's Guide.

AIRPORTS

Other than hard-surfaced runways — Seaplane Base

Hard-surfaced runways 1500 ft. to 8069 ft. in length

Hard-surfaced runways greater than 8069 ft. or some multiple runways less than 8069 ft.

Open dot within Hard-surfaced runway configuration indicates approximate VOR, VOR-DME, or VORTAC location.

All recognizable hard-surfaced runways, including those closed, are shown for visual identification. Airports may be public or private.

ADDITIONAL AIRPORT INFORMATION

Private "(Pvt)" – Non-public use having emergency or landmark value.

Military – Other than hard-surfaced. All military airports are identified by abbreviations AFB, NAS, AAF, etc. For complete airport information consult DOD FLIP.

Heliport-Selected Public — Unverified — Abandoned – paved, having landmark value, 3000 ft. or greater — Ultralight Flight Park Selected

Services–fuel available and field tended during normal working hours depicted by use of ticks around basic airport symbol. (Normal working hours are Mon thru Fri 10:00 A.M. to 4:00 P.M. local time.) Consult A/FD for service availability at airports with hard-surfaced runways greater than 8069 ft.

★ Rotating airport beacon in operation Sunset to Sunrise

AIRPORT DATA

Box indicates F.A.R. 93 — FSS — NO SVFR — F.A.R. 91 — Location Identifier — Airport Surveillance Radar — NAME (NAM) — CT – 118.3* ● ATIS 123.8 — 285 L 72 122.95 — UNICOM — VFR Advsy 125.0 — Airport of Entry

FSS – Flight Service Station

NO SVFR – Fixed-wing special VFR flight is prohibited.

CT – 118.3 – Control Tower (CT) – primary frequency

NFCT – Non-Federal Control Tower

* – Star indicates operation part-time (see tower frequencies tabulation for hours of operation).

● – Indicates Common Traffic Advisory Frequencies (CTAF)

ATIS 123.8 – Automatic Terminal Information Service

ASOS/AWOS 135.42 – Automated Surface Weather Observing Systems. NDB's broadcasting ASOS/AWOS data may not be located at the airport.*

UNICOM – Aeronautical advisory station

VFR Advsy – VFR Advisory Service shown where ATIS not available and frequency is other than primary CT frequency

285 – Elevation in feet

L – Lighting in operation Sunset to Sunrise

*L – Lighting limitations exist, refer to Airport/Facility Directory.

72 – Length of longest runway in hundreds of feet; usable length may be less.

When facility or information is lacking, the respective character is replaced by a dash. All lighting codes refer to runway lights. Lighted runway may not be the longest or lighted full length. All times are local.

RADIO AIDS TO NAVIGATION AND COMMUNICATION BOXES

VHF OMNI RANGE (VOR)

VORTAC

VOR-DME

Non-Directional Radiobeacon

RBn — POINT LOMA 302 — H+00 & ev 6m — Marine Radiobeacon

Other facilities, i.e., Commercial Broadcast Stations, FSS Outlets-RCO, etc.

122.1R 122.6 123.6 — OAKDALE — 382 *116.8 OAK

Underline indicates no voice on this freq

* – Operates less than continuous or On-Request.

T – TWEB ■ – HIWAS

R – Receive only

122.1R — MIAMI — Controlling FSS

122.1R — CHICAGO CHI

Heavy line box indicates Flight Service Station (FSS). Freqs. 121.5, 122.2, 243.0, and 255.4 (Canada – 121.5, 126.7 and 243.0) are normally available at all FSSs and are not shown above boxes. All other freqs. are shown.

For Local Airport Advisory use FSS freq. 123.6.

Frequencies above thin line box are remoted to NAVAID site. Other freqs. at controlling FSS may be available determined by altitude and terrain. Consult Airport/Facility Directory for complete information.

AIRPORT TRAFFIC SERVICE AND AIRSPACE INFORMATION

Only the controlled and reserved airspace effective below 18,000 ft. MSL are shown on this chart. All times are local.

Class B Airspace

Class C Airspace (Mode C See F.A.R. 91.215/AIM.)

Class D Airspace

40 — Ceiling of Class D Airspace in hundreds of feet. (A minus ceiling value indicates surface up to but not including that value.)

Class E Airspace

Class E Airspace with floor 700 ft. above surface

Class E Airspace with floor 1200 ft. or greater above surface that abuts Class G Airspace.

2400 MSL / 4500 MSL — Differentiates floors of Class E Airspace greater than 700 ft. above surface

Class E Airspace low altitude Federal Airways are indicated by center line.

Intersection – Arrows are directed towards facilities which establish intersection.

132° → V 69 — 169

Total mileage between NAVAID's on direct Airways.

Prohibited, Restricted, Warning and Alert Areas

Canadian Advisory and Restricted Areas

MOA – Military Operations Area

Special Airport Traffic Areas (See F.A.R. Part 93 for details.)

MODE C (See F.A.R. 91.215/AIM.)

National Security Area

Terminal Radar Service Area (TRSA)

IR211 — MTR – Military Training Routes

OBSTRUCTIONS

1000 ft. and higher AGL

below 1000 ft. AGL

Group Obstruction

Obstruction with high-intensity lights May operate part-time

2049 — Elevation of the top above mean sea level

(1149) — Height above ground

UC — Under construction or reported; position and elevation unverified

NOTICE: Guy wires may extend outward from structures.

MISCELLANEOUS

–1° E– Isogonic Line (1990 VALUE)

Ultralight Activity — Fl ★ Flashing Light

Hang Glider Activity — ● Marine Light

Glider Operations

NAME (Magenta, Blue, or Black) Visual Check Point

Parachute Jumping Area (See Airport/Facility Directory)

TOPOGRAPHIC INFORMATION

Roads

95 — 40 — Road Markers

Railroad

Bridges And Viaducts

Power Transmission Line

Aerial Cable

Landmark Feature – stadium, factory, school, golf course, etc.

Outdoor Theatre

Lookout Tower P-17 (Site Number) 618 (Elevation Base of Tower)

CG Coast Guard Station

Race Track

Tank–water, oil or gas

Oil Well ○ Water Well

Mines And Quarries

Mountain Pass

11823 (Elevation of Pass)

Rocks

Reef

Dams

Perennial Lake

Non-Perennial Lake

MATHEMATICS LABORATORY INVESTIGATION

AIRCRAFT NAVIGATION II

Topics: **ADDITION OF VECTORS**
Prerequisites: *Introduction to Vectors*
Equipment: IFR paper and laminated charts, protractor, compass or dividers, straight edge, dry erase markers

I. INTRODUCTION:

A student pilot must study charts and navigation in order to get a Private Pilot's License. Included in the requirements are both a written and a flight exam. These exams include chart reading. Primarily two charts are used - a sectional aeronautical chart used for *Visual Flight Rules (VFR)* and an enroute chart used for *Instrument Flight Rules (IFR).* IFR is used when weather and visibility are poor, during times of high air traffic, and for all commercial flights.

The region to become familiar with today is the Boston area airport, called Logan Airport. Included here for study is part of the "New York" IFR chart which includes the greater Boston area.

II. USING THE IFR (INSTRUMENT FLIGHT RULES) CHART:

For a pilot flying by instruments much of the information on the sectional chart obscures the data needed. Therefore a chart without topographical information is used. Since you can't see the ground when flying in poor weather conditions, much of that information would be unusable at that time anyway.

KEY IFR AIRWAYS

The IFR Chart highlights Vector Airways (Routes). Because these airways have both length (magnitude) and direction, they are *vectors*. The vector number is called the name of the route and always begins with a V.

You will find a *compass rose* located in several places on this chart. To orient yourself, locate the compass rose at Logan Airport. The arrow within the rose indicates the direction of north and vector directions are measured in a degrees from north in a clockwise rotation. Because of the prevailing winds around Boston, V-141 is the key route for Logan Airport. **What is the direction** of V-141?

Then look for the navigation aid , called a VOR, which

represents invisible intersections of vector routes in the air. The intersections, called intersects, are positioned over real navigation equipment on the ground.

A secondary intersect is symbolized by △ and is named with *made-up* words. These locations come from the intersection of radio beacons and are subject to triangulation error. (There is no navigation equipment below these marks.)

The direction of a vector route may change at either type of intersect unlike a vector which has only one direction associated with it.

VECTORS

As on the VFR chart, the initial vector direction is given near the beginning of the vector route. The number of nautical miles between intersects is given either above or below the vector while the cumulative mileage is given inside the symbol
The mileage in the box represents cumulative mileage between VOR's.

Once again **trace along V-141 leaving Boston**.

What is its initial direction?

Where does its direction change?

How many nautical miles to the location where the change in direction occurred?

Estimate the new compass direction.

Fill in the table below including all the vector airways departing Logan.

Vector Route Name	Initial Vector Direction	Vector Length to the first intersect	Vector Length to the first VOR
V-270-292	278°	37 nm	78 nm

III. PLANNING TO FLY A VECTOR ROUTE:

There is to be an air show outside of Bradley Airport. A pilot leaves Logan Airport at 9 A.M. for Bradley Aiport. Since he/she is the first plane on route, Air Traffic Control routes the pilot by the most direct routing, V1-419.

What is the initial direction for this vector?

Does this route change direction?

What is the total mileage for the trip?

If the plane is traveling at a speed of 135 knots *(nautical miles per hour)*, when will it be over Bradley?

A second pilot also leaves Logan five minutes later bound for Bradley. Since no two planes can be on the same vector route at the same time, this pilot is routed on V270-292 to GLYDE, V270 to SPENO, then V229 to DARTH, and finally V205 into Bradley.

In what direction and for how far did the pilot fly on each leg of his route?

Vector Route Name	Vector Direction	Vector Length to the next Intersect

If this plane was traveling at a speed of 145 knots, **when** would it be just outside of Worcester Municipal airport?

Which plane arrives in Bradley first?

Which plane travels the furthest?

Since each pilot left from the same Logan runway and arrives in the same place, the direct route can be called the ***resultant vector***. It is the equivalent result of any path that might have been taken.

Task 1: Draw a scaled representation of the flight of the two planes. Include and label the resultant vector.

Task 2: Rerouting for weather or emergency conditions.

At 10 A.M. a third plane is ready to take off for Bradley Airport when the Air Traffic Controller receives a message that there is a fire over Worcester and the smoke is billowing for a radius of 30 miles and that the wind is blowing Northeast. A pilot would need to prepare for many changes in plans. He/she needs to consider the changes in direction, mileage, ETA (estimated time of arrival), and fuel consumption. Assume the pilot has decided to fly at a speed of 145 knots.

What would a possible vector route be to avoid flying through the smoke?

If the plane burns 10 gallons per hour, **how much** fuel does it take to fly this longer route?

If the fuel price is $1.95 per gallon, **how much** extra does it cost for enough fuel to fly the avoidance route?

Task 3: Following a Flight Plan.

Boston Control gives a pilot the following Flight Plan:

Depart Logan on V141 and fly a constant speed of 140 knots.
Fly for 17 minutes to intersect and turn on a heading of 273^0.
Fly for another 20 minutes to the intersect and proceed from that VOR on a heading of 321^0 for the next 10 minutes.
Turn to heading 359^0. Fly 16 minutes.
Over the VOR turn on a heading of 111^0 and fly for 9 minutes.
Land at the nearest airport.

Fill in the details of each leg of the trip on the chart below.

Leg	Flying Time in minutes	Vector Direction	Vector Length in nm to the next Intersect	Vector Route Name	Name of the Intersect
Resultant					

What is the destination?

Using a straight edge and protractor **draw a vector representation** of the flight plan. Make your graph as accurate as possible. Label each leg of the route. **Highlight and label the resultant.**

Task 4: Wind Effect

A plane will be "pushed" sideways by wind blowing across the plane's heading. The ground track is the resultant vector of the wind and the plane's heading.

If you plan to fly from Logan Airport to Norwich Airport at a speed of 150 knots on the most direct vector and there is a 20 knot wind from the South, you will need to decide on the compass heading.

Draw a vector representation of the situation, using the same directions as in Task 3.

State the magnitude and direction of the resultant vector.

APPENDIX: The Maneuvering Board

Navigators use vectors to represent the speed and direction of planes (or ships). The direction of the plane is determined from magnetic North. Today's navigators seldom use the compass points, North, South, East, etc. They instead use a 360 degree compass where North is represented by 0^0 (or 360^0), East by 90^0, South by 180^0, and West is 270^0.

The Maneuvering Board, called by navigators a "MO Board", represents this compass. This special graph paper is published by the US Defense Mapping Agency as its origins lie with military applications. Navigators use the Maneuvering Board to add vectors.

Let's use a sample problem to illustrate the use of the board. A plane leaves from Logan traveling along vector V3, whose heading is 030 and magnitude (length) is 14 NM. Find the direction 030 on the outside circumference of the sample chart. Draw a light line from the center of the chart through the 030 mark on the circumference. This is the first vector's direction. The length of this vector must be scaled. Note the 4:1 scale on the right side of the Board. This will be an appropriate scale for the sample problem. The 14 nautical mile length is set by opening your compass or divider along the 4:1 scale and that length should be transferred to the Board direction line you have just drawn. Draw an arrowhead at the end of your vector to signify direction.

Since adding vectors is done by placing tail to tip, we must draw the vector representing the next leg of the trip starting from the tip (arrowhead) of the just drawn vector. For the second leg of this sample trip the plane changes heading to 021 and travels 24 nm to Pease Air Force Base. The 021 heading is carefully transferred by protractor or parallels to the tip of the first vector. Again draw a light line to indicate direction. To determine the vector length, remember that you are using a 4:1 or one-fourth scale. The Resultant of these two vectors, the shortest distance from Logan to Pease is found by drawing a vector starting from the center of the Board to the tip of your last vector. For this sample, the Resultant is: heading 024.5 for a distance of 38.5 nm.

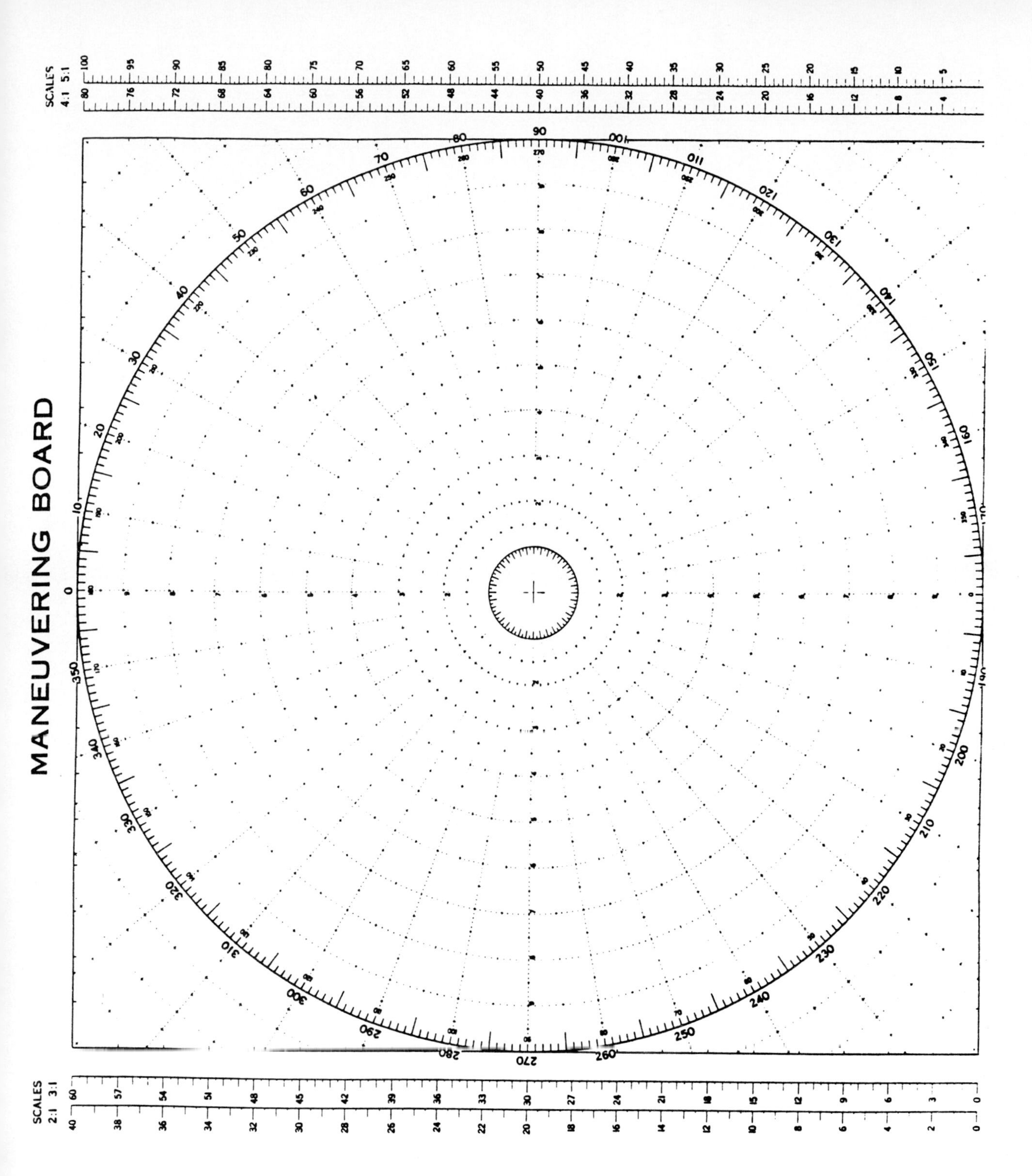
MANEUVERING BOARD
SCALES
4:1 5:1
SCALES
2:1 3:1

MATHEMATICS LABORATORY INVESTIGATION

AIR POLLUTION

Topic: **Normal Distributions, Standard Deviation, Logarithmic Plots**
Prerequisite: *Exponential Functions*

I. INTRODUCTION:

Among the greatest drawbacks to the Industrial Revolution has been the associated proliferation of smokestacks that continuously spew forth the gaseous and particulate pollutants (called *detritus*) of manufacturing processes. The smoke from these stacks (the *plume*) contains all manner of dust, dirt, and vaporized pollutants which individually, and in concert with each other, are capable of causing serious adverse health effects on humans. Indeed, if the concentrations are high enough, the emitted pollutants can kill people.

It would be useful, therefore, to be able to calculate, prior to construction , what the probable concentration of pollutants will be
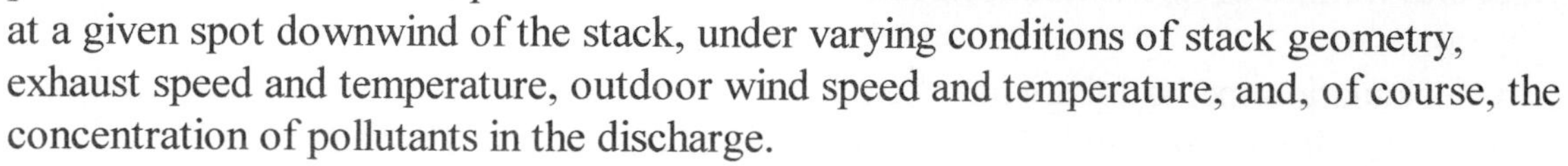
at a given spot downwind of the stack, under varying conditions of stack geometry, exhaust speed and temperature, outdoor wind speed and temperature, and, of course, the concentration of pollutants in the discharge.

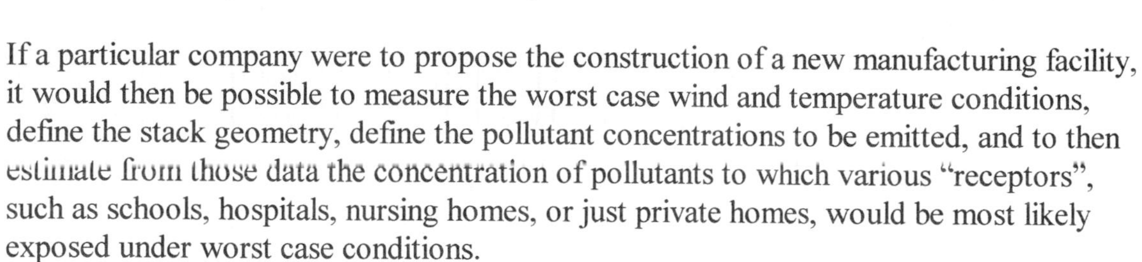

If a particular company were to propose the construction of a new manufacturing facility, it would then be possible to measure the worst case wind and temperature conditions, define the stack geometry, define the pollutant concentrations to be emitted, and to then estimate from those data the concentration of pollutants to which various "receptors", such as schools, hospitals, nursing homes, or just private homes, would be most likely exposed under worst case conditions.

If "worst case" is defined to mean the conditions under which the maximum concentration of pollutants will affect the "receptors", the worst case concentration can be calculated and compared to a "standard" or acceptable concentration. From that comparison, it can be determined whether the expected concentration will exceed the standard and, therefore, be unsafe. If that happens, a "safe" concentration, typically taken to mean the "standard" concentration, can be established and the calculations can be run backwards to determine the allowable discharge concentration under the worst case conditions. That concentration then becomes the maximum concentration which the facility is allowed to discharge under any conditions.

It is also then possible to adjust the calculations a bit to account for the concentration at a receptor which may not be directly downwind of the stack, but which still may be affected by the emissions because of the tendency for pollutants to spread out in a fan-like shape as they leave the stack. The fan shape is both horizontal and vertical. Therefore, a receptor off to the side of the main plume flow line may still be affected. Since there is necessarily more dilution of the pollutants in order for them to reach a point off to the side of the main line of flow, there will never be a problem at a side receptor if there is no problem along the main line of flow, unless the allowable concentration is significantly lower at that side receptor.

If, however, the side line receptor requires a lower allowable concentration than the main line receptor, it is useful to know what concentration will reach that side line receptor and to be able to calculate back from that allowable concentration to an allowable stack discharge concentration.

Moreover, it is useful to be able to vary the different parameters affecting the concentration of pollutants at the receptor location to see what effect unusual wind or weather conditions might have, or to see what effect changing the stack geometry of stack discharge parameters might have on the pollutant concentration at the receptor locations. This approach allows the industry to build its facility, while ensuring protection of the health and welfare of downwind residents.

Such a calculation methodology has been developed (due to D. B. Turner, [Tur]) and the use of the resulting equation is the subject of this investigation. That equation is:

$$\chi(x,y,H)=\left(\frac{Q}{\pi\sigma_y\sigma_z u}\right)\exp\left(-\frac{1}{2}\left(\frac{y}{\sigma_y}\right)^2\right)\exp\left(-\frac{1}{2}\left(\frac{H}{\sigma_z}\right)^2\right) \qquad \text{(Eq. 1)}$$

where x is the distance downwind (in meters),
y is the "crosswind" distance (in meters),
H is the effective stack height (in meters), to be defined shortly,
Q is the emissions rate of the pollutants (in grams/sec),
u is the wind speed (in m/sec),
σ_y and σ_z are the plume standard deviations (in meters), to be defined shortly,
and $\chi(x,y,H)$ is the concentration of pollutants at point (x,y) (in g/m^3)

(Here, exp(x) is the exponential function, e^x).

EFFECTIVE STACK HEIGHT

The height of the stack needs to be adjusted for both the stack geometry and weather conditions. For example, if the temperature of the stack emission is much higher than the air temperature, the pollutants will rise quickly into the air, effectively giving the same results as cooler pollutants being emitted from a taller stack. This adjusted stack height is called the effective stack height and can be computed by the following formula (due to J. Z. Holland, [Hol]).

$$H = h + \frac{v_s d}{u}\left[1.5 + \left(.0268P\left(\frac{T_s - T_a}{T_s}\right)d\right)\right] \qquad \text{(Eq. 2)}$$

where, for the stack geometry,

h is the height of the stack (in meters),

d is the diameter of the stack (m),

v_s is the stack velocity (m/sec),

T_s is the temperature of the pollutants leaving the stack (°K),

and, for the weather conditions,

u is the wind speed (m/sec),

T_a is the air temperature (°K),

P is the air pressure (kPa).

PLUME STANDARD DEVIATIONS

The plume standard deviations (or *horizontal* and *vertical dispersion coefficients*) are measures of how wide the plume is in the horizontal (crosswind) and vertical directions. These are functions of the downwind distance (x) and the stability of the atmosphere and can be computed graphically from the attached tables. Table 1, entitled **Key To Stability Categories** categorizes atmospheric stablility into six classes (A-F) depending on wind speed, whether it is day or night, amount of overcast and amount of solar radiation. Tables 2 and 3 compute the plume standard deviations from the atmospheric conditions and the distance downwind. (Note that in these tables, the distance downwind is measured in km. In all other equations, it is in meters.)

Note: Tables 2 and 3 are "log-log" graphs. When reading, for example, the x-axis between 1 and 10, there are 10 grid lines per unit between 1 and 5 but five grid lines per unit between 5 and 10.

For a graphical interpretation of some of these variables, see Figure 1 below.

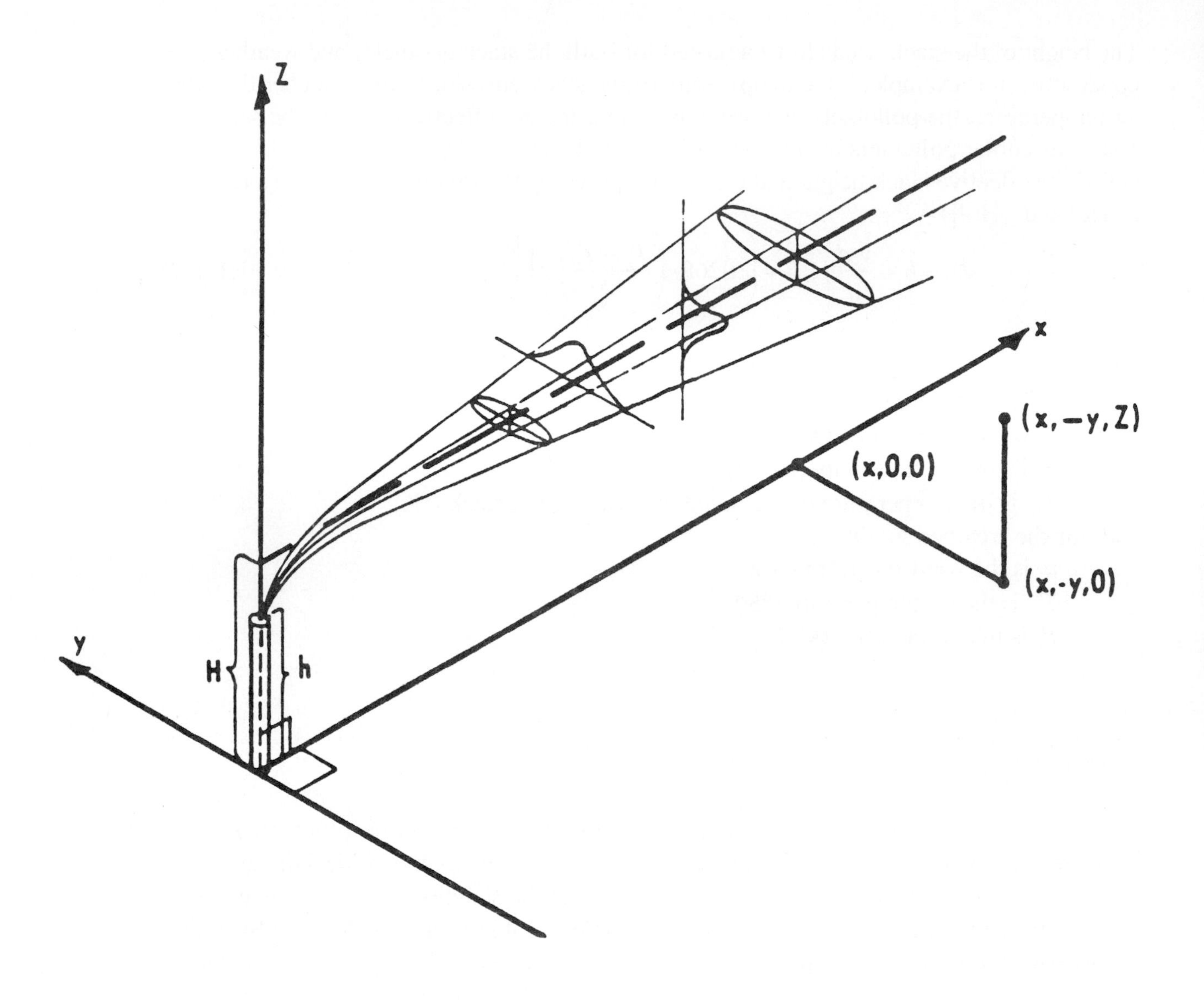

Figure 1: Coordinate System Showing Horizontal and Vertical Distributions
([Tur], p. 5)

II: SITUATION:

The XYZ Corporation wants to build a manufacturing facility at the location shown on the attached map. The most significant worry is the maximum concentration of pollutants which might occur at the elementary school, the location of which is also indicated on the map.

Task 1: Examining the Pollution Levels at the School

Part A: Given the prevailing wind direction shown on the map and the scale of the map, determine *x*, the distance downwind from the stack of the elementary school and *y*, the crosswind distance from the stack of the school.

Part B: Given the following stack parameters and weather conditions, what is the maximum expected concentration of pollutants at the elementary school if the emissions rate of the pollutants in the exhaust gas is 30,000 g/sec? (From Equations 1 and 2. BEWARE OF UNITS.)

Stack Parameters		Weather Conditions	
Stack Height	100 ft	Wind Velocity	40 mph
Stack Gas Exit Velocity	85 ft/sec	Air Temperature	85° F
Stack Gas Exit Temperature	120° F	Air Pressure	95 kPa
Stack Diameter	5 ft	Stability Category	C

Task 2: Varying the Conditions

Part A: If it were determined that the maximum allowable concentration of pollutants at the school is 20 mg/m^3, what is the maximum pollutant emissions rate allowable at the stack (keeping all other parameters the same as above)?

Part B: The XYZ Corporation decided that raising the stack height to 200 feet would probably dilute the pollutants enough to allow them to discharge the pollutants into the atmosphere at a higher rate, without exceeding the 20 mg/m^3 standard at the school. Calculate the change in the pollutant emissions rate allowable from the stack if the stack height is raised to 200 feet, but all other parameters remain constant.

Task 3: Examining the Crosswind Distribution of Pollutants

Part A: In order to understand the distribution of pollutants, it helps to break Equation 1 apart and analyze the terms individually. For a fixed distance, x, downwind, we can define A_x to be the *total amount of pollution* x meters downwind. It would be computed by the following formula:

$$A_x = \frac{Q}{u\sigma_z\sqrt{\pi}} \exp\left(-\frac{1}{2}\left(\frac{H}{\sigma_z}\right)^2\right) \qquad \text{(Eq. 3)}$$

Using x as computed in Task 1 for the elementary school, A_x would be the total concentration of pollutants at all points x meters downwind of the stack (regardless of their crosswind distance from the main line of flow). Compute A_x for x as computed in Task 1.

Part B: To see how this pollution gets distributed for different values of y, one can use the formula

$$\chi(x, y, H) = \frac{A_x}{\sigma_y\sqrt{\pi}} \exp\left(-\frac{1}{2}\left(\frac{y}{\sigma_y}\right)^2\right) \qquad \text{(Eq. 4)}$$

Considering x (the downwind distance of the elementary school) to be a constant, A_x and σ_y then become constants as well. Equation 4 is then a function of a single variable, y. Use a graphing calculator (or computer software) to graph this function.

Part C: The function graphed in Part B is called a *Normal (or Gaussian) Distribution.* (Actually, it is a constant, A_x, times a normal distribution, but that is a minor point here.) Geometrically, it is the "Bell Shaped Curve" we think of with respect to normally distributed data. The quantity σ_y is the *standard deviation* of this distribution. A fact from Statistics about normal distributions is that about 68% of the data lies within one standard deviation of the mean (in this case $y = 0$) and 95% of the data lies within two standard deviations of the mean. For this problem, this translates to the statement that 95% of the pollution lies within $2\sigma_y$ meters of the main line of downwind pollution. In this model, σ_y measures the width of the plume of pollutants. Using Table 2 (Horizontal Dispersion Coefficient), explain how the width of the plume changes as you move downwind (fix a stability category). Then, for a fixed downwind distance, explain (from the same table) how varying the weather conditions varies the width of the plume.

Task 4: Examining the Downwind Distribution of Pollutants

The function A_x from the previous task is the distribution function for the downwind distribution of pollutants. This is a function of the downwind distance, x. Unfortunately, it is not easy to analyze since we do not have a formula for the vertical dispersion coefficient, σ_z.

Part A: Using a ruler (**not the grid lines on the graph**), compute the slope of the line on the Table 3 (Vertical Dispersion Coefficient) corresponding to Stability Category C. Call this slope m.

Part B: Since Table 3 is a "log-log" graph, having slope m means that

$$\log \sigma_z = m \log\left(\frac{x}{1000}\right) + k$$

(Remember that this table measures downwind distance in kilometers, hence the 1000 in the formula) or, equivalently,

$$\sigma_z = C\left(\frac{x}{1000}\right)^m$$

where C ($=10^k$) is the value of σ_z at $x = 1000$. Using this formula in Equation 3 and a graphing calculator (or computer software), graph the downwind distribution, A_x, in terms of x, keeping the parameters as previously defined. (Since σ_z is *not* a constant, this will not be a normal distribution and will not look like a "Bell Shaped Curve".) From this graph, estimate how far downwind will be the highest concentration of pollutants.

Part C: How does changing the stack height to 200 feet (as in Part B of Task 2) affect the graph in Part B?

EXTENSIONS:

1. How does changing the weather conditions affect the Effective Stack Height (Equation 2) and what affect does this have on the graph in Part B of Task 5?
2. Use software to graph Equation 1 as a function of x and y. (This will mean modelling both dispersion coefficients.) Examine the resulting surface and/or contour plot.
3. See if you can find the tables listed above (or the references listed below) at an Environmental Protection Agency Web Site.

Table 1

Key to Stability Categories ([Tur] p. 6)

Surface Wind Speed (at 10 m) m/sec	Day			Night	
	Incoming Solar Radiation				
	Strong	Moderate	Slight	Thinly Overcast or ≥ 4/8 Low Cloud	≤ 3/8 Cloud
< 2	A	A-B	B		
2 - 3	A-B	B	C	E	F
3 - 5	B	B-C	C	D	E
5 - 6	C	C-D	D	D	D
> 6	C	D	D	D	D

Notes:

1. The neutral class, D, should be assumed for overcast conditions during day or night.
2. Night refers to the period from one hour before sunset to one hour after sunrise.
3. "Strong" incoming solar radiation refers to solar altitude greater than 60° with clear skies; "slight" refers to a solar altitude from 15° to 35° with clear skies. Incoming radiation that would be strong with clear skies can be expected to be reduced to moderate with broken (5/8 to 7/8 cloud cover) middle clouds and to slight with broken low clouds.

Table 2

Horizontal Dispersion Coefficient as a Function of Downwind Distance ([Tur] p. 8)

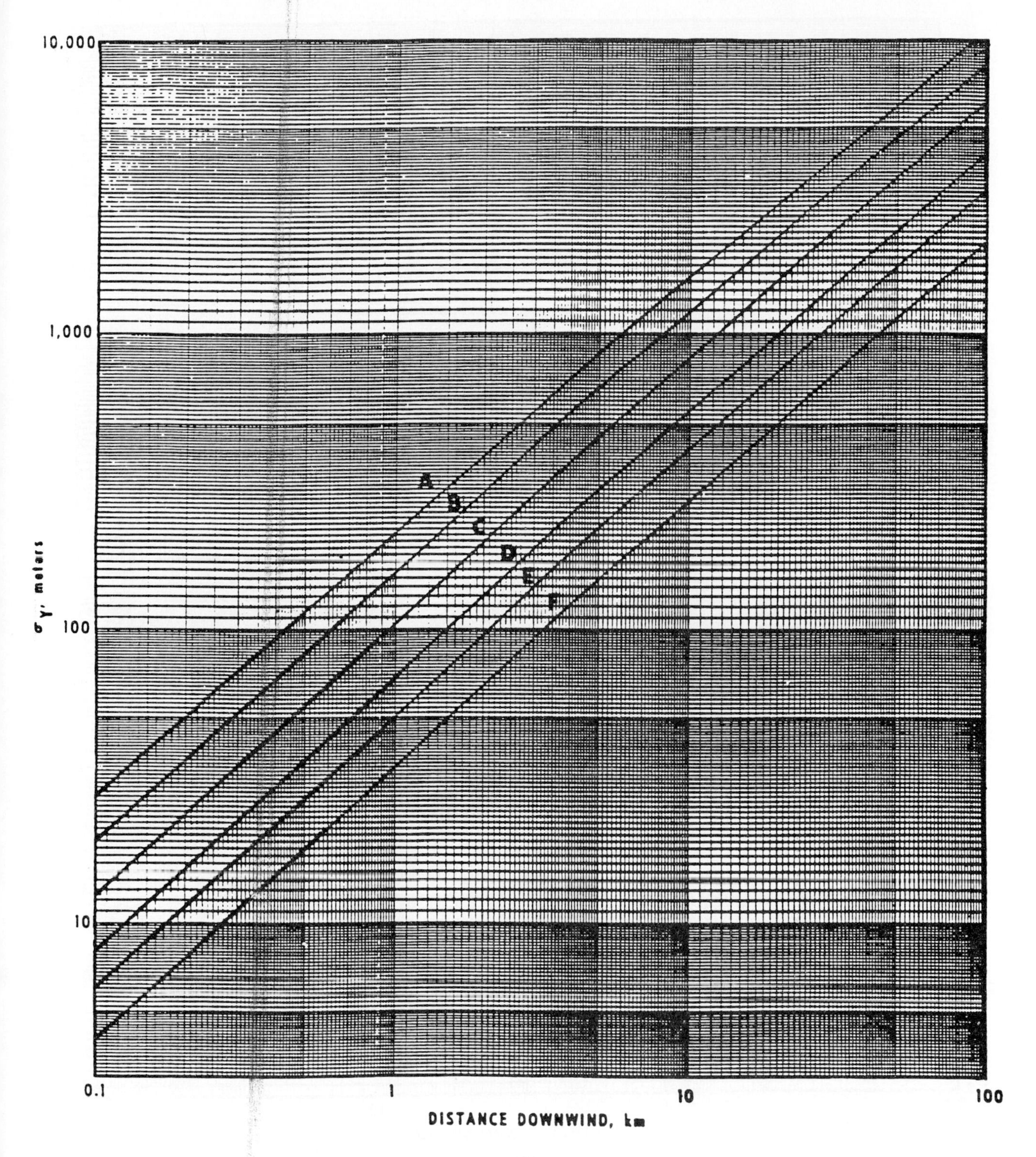

11

Table 3

Vertical Dispersion Coefficient as a Function of Downwind Distance ([Tur] p. 9)

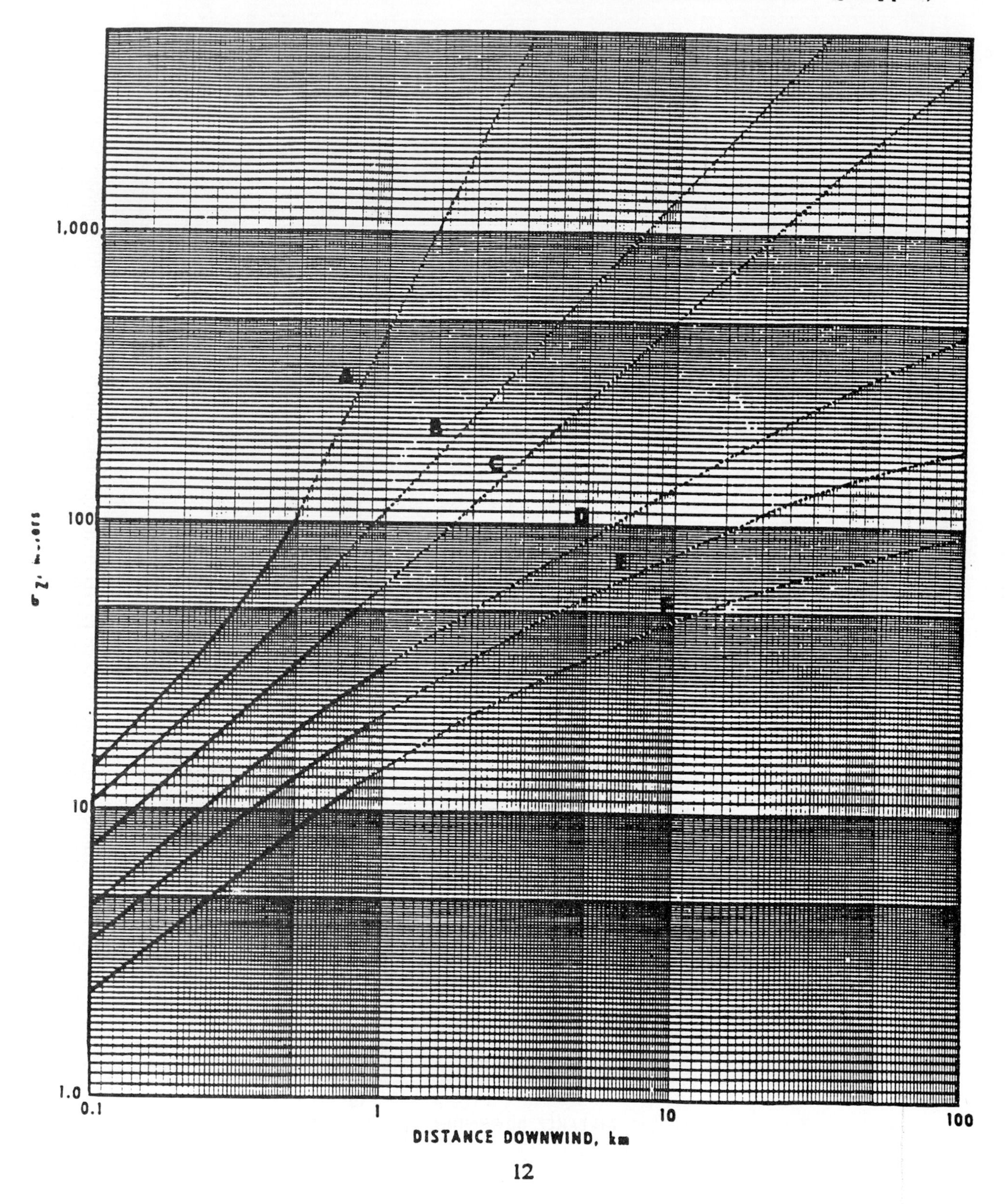

12

REFERENCES:

[Hol] J. Z. Holland, *A Meteorological Survey of the Oak Ridge Area* (U.S. Atomic Energy Commission Report No, ORO-99), Washington D.C: U. S. Government Printing Office, 1953, p. 40.

[Tur] D. Bruce Turner, Workbook on Atmospheric Dispersion Estimates (U.S. Department of Health, Education and Welfare, Public Health Service, National Center for Air Pollution Control Publication No. 999-AP-28)

MATHEMATICS LABORATORY INVESTIGATION

BUCKLING OF SLENDER COLUMNS

Topic: **VARIATION - INVERSE SQUARE**
Prerequisite knowledge: *Plotting data, functional notation*

I. INTRODUCTION

While we most often think of columns as components in an architectural design or in an underlying structure, there are many examples of columns in the natural world. Consider a tree after an ice storm, the weight of the snow and ice may cause the tree to bend or *buckle*. As the temperature rises the ice and snow melt and the tree returns to its original upright position. The tree is exhibiting *elastic buckling*. However, if the weight of the ice and snow is too great, the tree will snap. This is a sudden and irreparable event.

Photo courtesy of Bell Athletics, Inc.

The pole used by a pole-vaulter exhibits the same phenomenon. As the athlete ascends, the pole bends. When the weight of the athlete is released the pole returns to its original straight form (elastic buckling). If the pole were to break, it would be sudden and irreversible.

The effect of buckling on the columns of a building, bridge, or any other structure is similarly critical. Architects and engineers must know what load a column can support. Initially when a vertical load is applied to a column, the column compresses. For a short thick column a heavy enough load will cause the column to failed in compression. (A crushed soda can is an example of a failed short column.) A tall slender column, on the other hand, will buckle before it fails in compression. The load which causes elastic buckling is called the **Euler Load** (pronounced 'Oiler').

II. BACKGROUND

The load that a column can support depends on many factors. Certainly the type of material that is used will make a difference in the capacity of the column. The *Modulus of Elasticity* (denoted E) is a measure of the strength of the material. (See Mathematics Laboratory Investigation: Strength of Materials.) The shape of the column and the direction in which it bends also contribute to its capacity. The *Moment of Inertia* of a cross-section of the column (denoted I) is used to measure the effect of the shape of the column. This investigation utilizes only one type of material and all of the columns have the same cross-section so E and I will be considered constant.

There is one other factor which must be considered: the *effective length* of the column (denoted L).

III. LENGTH OF COLUMN

Before you start to collect data, get the "feel" of your columns by gently applying vertical pressure to them. This investigation is concerned with elastic buckling so we don't want the columns to break. Remember that failure do to buckling is sudden and irreversible. (Stand a soda straw or coffee stirrer on end and apply vertical pressure. You'll see that it will buckle, but suddenly it folds and fails completely. It's all over!)

TASK 1: Collecting data.

Record the lengths of each of the columns that you are supplied with. Using the bathroom scale, find and record the vertical load required to buckle each one of the columns. Be very careful not to break the columns. We only want a measure of the load that is necessary to cause buckling.

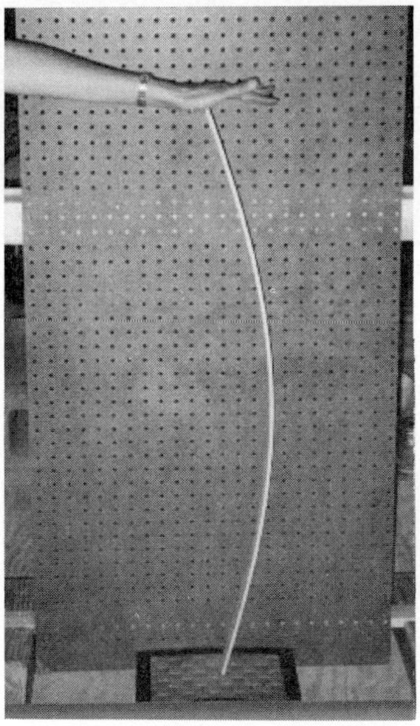

Take each measurement repeatedly until you are confident in your data. Good data collection requires cooperation among your group members. Each group member should get the feel for when buckling occurs but, for consistency, one member should apply the forces when you are gathering data.

Length, L	**Euler Load, P(L)**

TASK 2: Plotting data.

From the data in your table, construct a scatterplot of Length vs. Load. Be sure to label the axes including the units of measure.

Sketch in a smooth curve that best fits the data. *This curve represents the Euler Load as a function of length.*

At this point, you could enter the data into your calculator and do a power regression on it. What equation best fits the data?

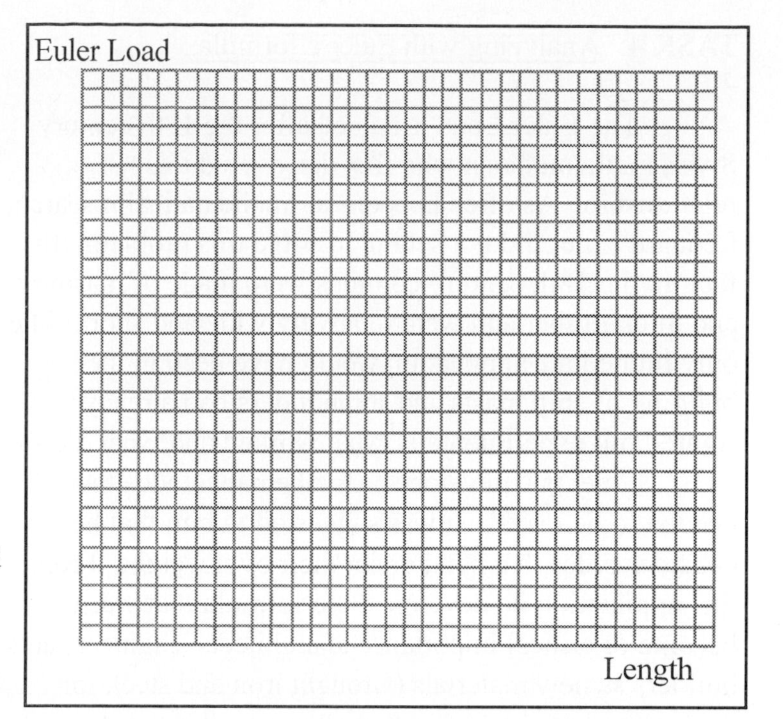

TASK 3: Consider your results.

A. What load would theoretically be required to buckle an extremely long ($L \rightarrow \infty$) column made of the same material as the columns that you just used? Why?

What characteristics of the graph reflect your answer?

B. What load would theoretically be required to buckle an extremely short ($L \rightarrow 0$) column of the same material? Why?

What characteristics of the graph reflect that tendency?

Do you think that this conclusion is realistic? Why or why not?

TASK 4: Analyzing with Euler's formula.

The Euler Load is named after the 18th century Swiss mathematician who first discovered this relationship. As often happens in mathematical research, Léonard Euler did not set out to solve this problem. In fact, in the 18th century columns were made of masonry and thick timber and were relatively wide and short. The only immediate application would have been in the building of ship masts and shipbuilders had already figured out experimentally what worked and what didn't.

Euler had developed a mathematical idea that needed to be tested. It was suggested to him that he consider poles that bend under their own weight. From this application he developed the formula that was to become of critical importance to architects, engineers and builders as new materials (wrought iron and steel, for example) were developed that allowed tall slender columns to be built.

Euler's Formula

The load "P" which causes elastic buckling is given by the formula $P = \frac{\pi^2 E\, I}{L^2}$

where P is the Euler Load, E is the Modulus of Elasticity, I is the Moment of Inertia of a cross section and L is the effective length of the column.

In this part of the investigation we are considering the Euler Load as a function of length. In order to do this, E and I are held constant (the same material is used for all of the columns, the cross sections of the columns are the same and the columns are all bending in the same direction), so P and L are the only variables that are currently being considered. The formula can be written in functional notation:

$$P(L) = \frac{\pi^2 EI}{L^2}$$

Remember that P(L) represents the value of P resulting from the value of L, not P multiplied by L.

The formula and some algebra can be used to predict the effect of changing the length of the column. E.g. If the length of the column is cut in half we replace L with one-half of L:

$$P(0.5L) = \frac{\pi^2 E\, I}{(0.5L)^2} = \frac{\pi^2 E\, I}{0.25\, L^2} = 4 \cdot \left(\frac{\pi^2 E\, I}{L^2}\right)$$ or 4 times the original Euler Load.

The shorter column can bear a load that is **4** times as great. In mathematics this type of relationship is called an ***inverse square relationship***.

What would happen if the column were doubled in length?

What effect would tripling the length of the original column have on the Euler Load?

What if the column were only 40% of its original length? Could the column support more or less than the original load? By how much?

How does your graph from Task 2 fit Euler's formula?

IV. EFFECTIVE LENGTH

When bracing is present the length to be considered is no longer the actual length of the entire column.

TASK 5: Adding bracing.

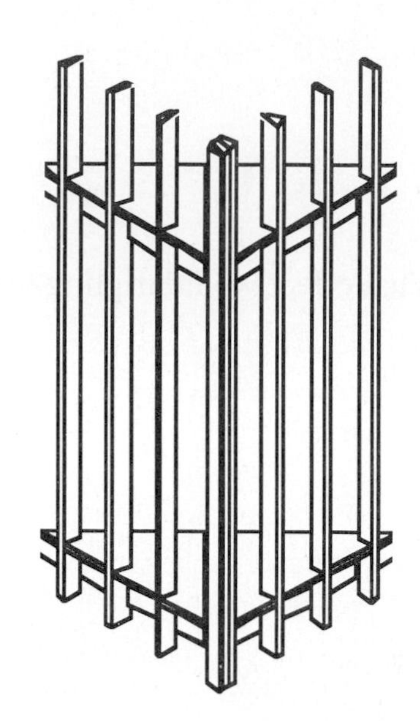

In a building the bracing takes different forms. The sketch on the right shows an example in which the bracing is provided by notches in the floor joists.

Choose one of your longer columns and have one of your group members place his/her thumb and forefinger on either side of the center point of the column to provide the bracing at that point. The goal is to prevent the midpoint of the column from moving to the left or right without preventing it from moving vertically. Apply vertical pressure and record the load required to buckle the column. (This will take some practice.)

What shape did the deflection curve take?

What load was necessary to cause buckling?

How does this load and deflection curve compare to the results for same column without bracing?

Compare this load and deflection curve to the ones obtained by using an unbraced column half as long.

When bracing is present, the length that is used in Euler's Formula is no longer the actual length of the column. Instead the *effective length* is used. In general, the effective length is the distance between braces.

Experiment with placing two braces on a column. Sketch the shape of the column after it has buckled and give its effective length.

Column Length:__________

Effective Length:_________

A typical building has stories that measure 12 to 15 ft high. Since flatbed trucks can legally carry 40 ft long loads without special permits, columns are often used that measure 30 to 40 ft long. If a 2 story building is built with 30 ft long columns, what is the effective length of these columns?

Why would a designer choose one long column instead of 2 shorter ones?

On February 26, 1993 6 people were killed and 1000 were injured when a car bomb in the garage below the World Trade Center blew out some of the concrete floors as shown in the sketch. How did the effective length of the columns change? What was the resulting change in capacity of the columns?

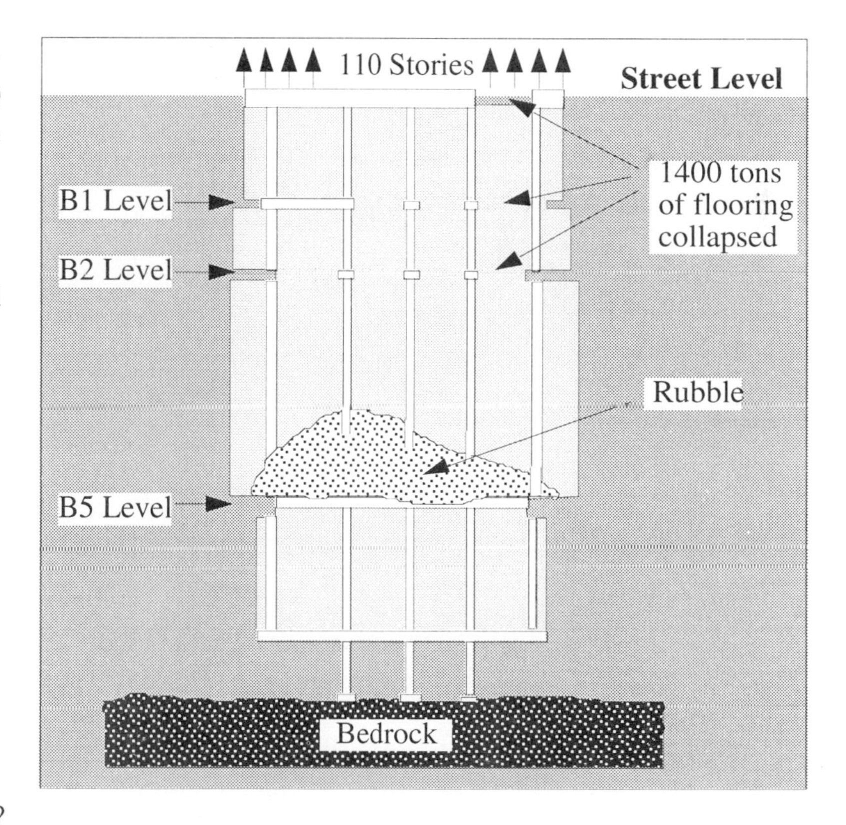

When stabilizing the building, one of the engineers' first concerns was to add bracing to the columns. Why?

TASK 6: Bracing at the base of the column - a special case.

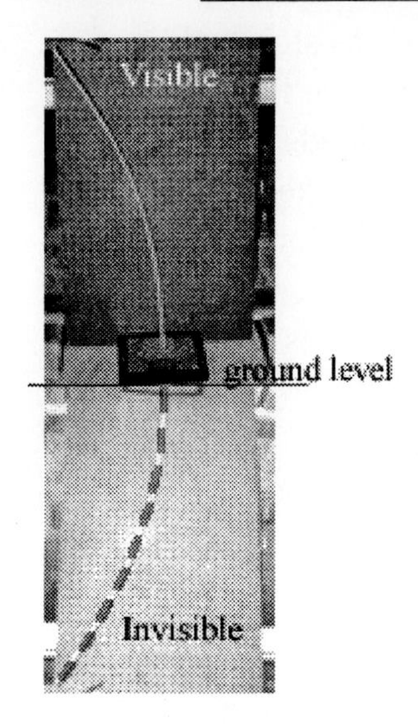

There are many examples of tall slender columns that are braced only at their bases: flag poles, lights for a playing field, tall trees, smoke stacks, ... In this case only half of the effective length of the column is visible. While the entire actual column is visible, there is an imaginary extension of the column going into the ground.

Choose one of your longer columns and add bracing at the base of the column. As before, we just want to prevent the column from moving at the base; the base of the column should remain perpendicular to the scale. When you apply a vertical force this case will feel very different. Think of it as resting your hand on the column. Let it move sideways and down naturally.

Write a paragraph or two in which you discuss the answers to the following questions: What is the shape of the deflection curve? How does it differ from the curve you saw in Task 1? How is it similar? What is the Euler Load for the column braced in this way? How does it compare to the Euler Load for the same column without bracing? What is the effective length for the column?

TASK 7: Summary.

Fill in the table below for a steel column that is 30 ft (how many inches?) long, with a rectangular cross section which gives a Moment of Inertia of 48 in^4. The average value for the Modulus of Elasticity of steel is 30 x 10^6 pounds per square inch.

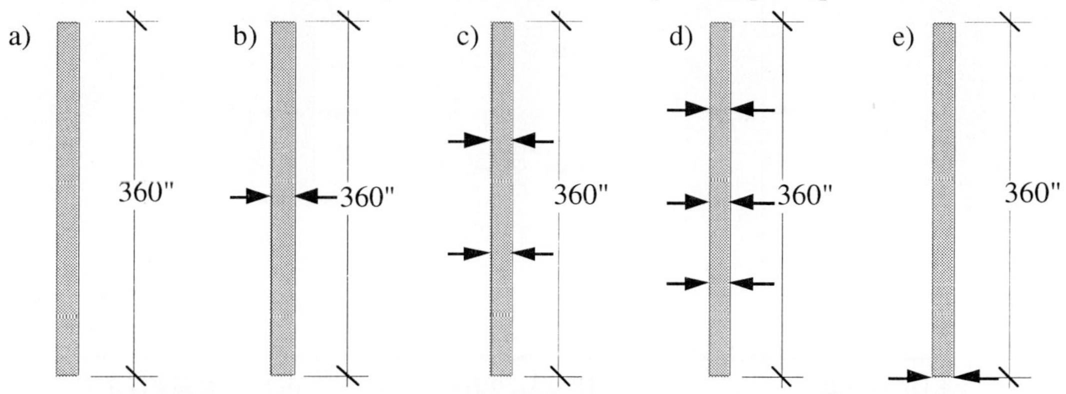

Bracing	Effective Length, L	Euler Load, P(L)
a) None		
b) At midpoint		
c) At thirds		
d) At fourths		
e) At the bottom		

Generalize by considering a column of length L to which bracing is added. Fill in the chart with the values for effective length in terms of L and Euler Load in terms of P.

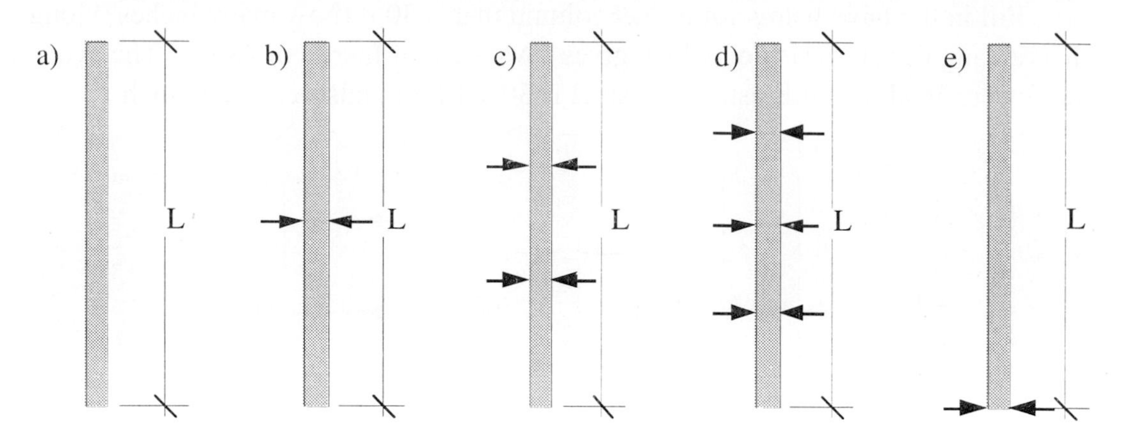

Bracing	**Effective Length, L**	**Euler Load, P(L)**
a) None	L	P
b) At midpoint	L/2	4P
c) At thirds		
d) At fourths		
e) At the bottom		

What is the relationship between the effective length of a column and its' Euler Load? Why is it considered mathematically to be an "inverse square" relationship?

TASK 8: <u>Generalize to other fields.</u>

The inverse-square relationship appears in formulas in many different fields. The following problems are just a sample.

A. The illumination level from a light varies inversely with the square of the distance from the light source. This relationship is written as

$$E = \frac{cp}{r^2}$$

where E is the illumination level (in foot candles)
cp is the candlepower of the course of light (in candles) and
r is the distance from the light source (in feet).

If you move from a position 25 ft from a spotlight to a position 50 ft from it, what change in illumination will you experience?

B. Newton's universal law of gravitation states that the force F of gravitation between two
objects varies jointly as the masses m_1 and m_2 of the objects and inversely as the square of the distance r between their centers.

$$F = \frac{Gm_1m_2}{r^2}$$

If the distance between two objects were cut in half, what would be the effect on the gravitational force between them?

C. The electrical resistance R of a wire varies inversely with the square of the diameter of the wire.

$$R = \frac{(\alpha + \alpha t)L}{d^2}$$

If the diameter of the wire was increased by 75%, what would the resulting change in the resistance be?

D. The total resistance of two resistors R_1 and R_2 connected in series varies inversely with the current I.

$$R_1 + R_2 = \frac{P}{I^2}$$

If the current was decreased by 10%, how would the total resistance change?

E. The length of time required to empty a tank varies inversely with the square of the diameter of the drain pipe.

If the pipe were clogged so that its effective diameter were only one-third of the original diameter, how much longer would it take to drain the tank?

F. The power gain by a parabolic microwave dish varies directly as the square of the diameter of the opening and inversely as the square of the wavelength of the wave carrier.

What percentage gain in power would result from a 30% decrease in the wavelength of the wave carrier?

EXTENSION TO BUCKLING OF SLENDER COLUMNS
MOMENT OF INERTIA

Did you notice that all of the columns buckled in the same direction, parallel to the shortest side of the cross section of the column? Euler's formula predicts the load at which elastic buckling will occur, therefore the smallest load that causes buckling is the Euler Load. Consequently when there is a choice, the smallest value for the Moment of Inertia will be used and will indicate the axis along which buckling will occur.

TASK 1: <u>Calculating the Moment of Inertia (I).</u>

The Moment of Inertia depends on the geometry of the cross section of the column. Sketch a cross section of one of the columns placing the x and y axes as shown. Include the actual dimensions of the cross section of your columns in your sketch.

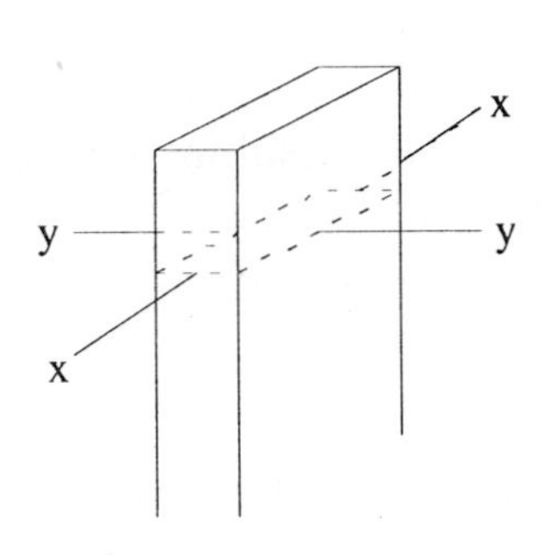

Cross Section:

In a rectangle one of the sides is called the base, b, and the other side is called the height, h. Many formulas from geometry do not specify which of the sides to call the base and which to call the height. However the formula for the Moment of Inertia <u>always assigns h to the dimension perpendicular to the axis that is being considered</u>.

So, to find the *Moment of Inertia about the x-axis* (I_x)of the column under consideration:

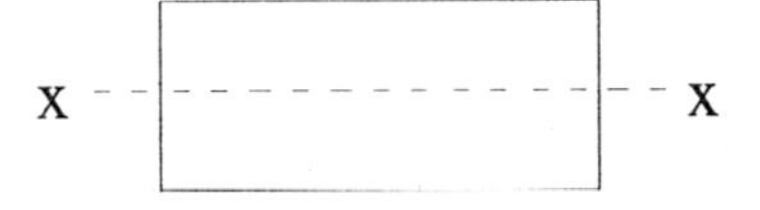

the measurement of the height is h = _______

the measurement of the base is b = ________

To find the *Moment of Inertia about the y-axis* (I_y):

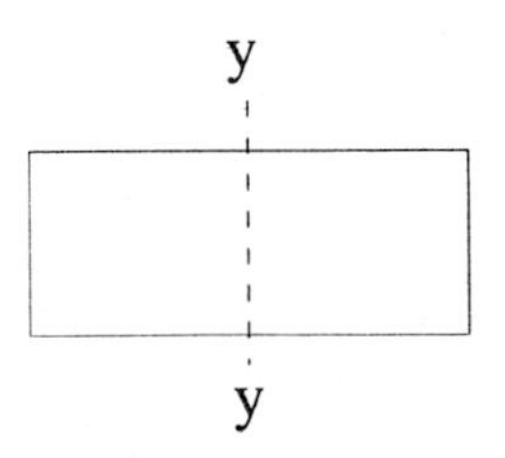

the measurement of the height is h = _______

the measurement of the base is b = ________

$I = \frac{bh^3}{12}$ Use the formula for the moment of inertia to calculate both I_x and I_y paying special attention to the values of b and h for each.

What is the value for I_x and I_y? Be sure to include units in your calculations and answers.

$I_x =$ ______________

$I_y =$ ______________

Which of these values for I should be used in calculating the Euler Load for this column? Why?

If you have a set of equal length columns with different rectangular cross sections, use the scale and experimentally find the Euler Load for each of the columns. If not, continue with Task 6.

TASK 3: Collecting data.

Calculate the value of I for each of the columns. Measure and record the Euler Load required to buckle each of them. (Remember that the Euler Load occurs with the minimum Moment of Inertia for the cross section of the column.)

Base of Cross Section	**Height of Cross Section**	**Moment of Inertia, I**	**Euler Load, P(I)**

TASK 4: Plotting data.

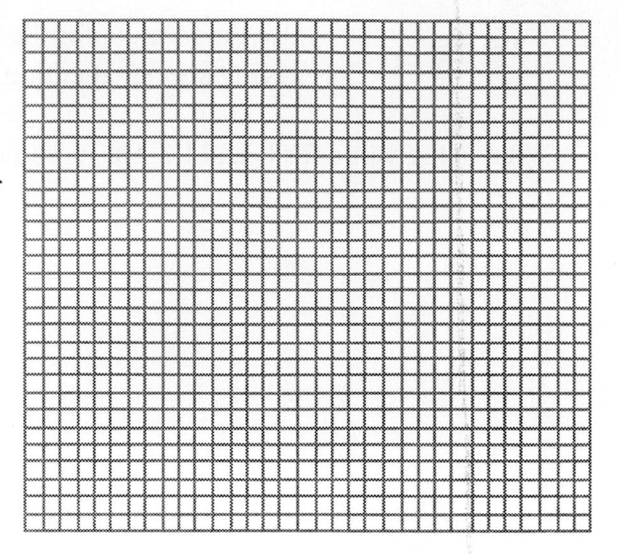

From the data in your table, construct a scatterplot of Moment of Inertia vs. Load. Be sure to label the axes including the units of measure. Notice that the data falls approximately along a straight line.

Sketch in a line that best fits the data. This line represents the Euler Load as a function of the Moment of Inertia.

TASK 5: Consider your results.

Why is this graph so different from the graph in Task 2 of **Mathematics Laboratory Investigation Buckling of Slender Columns**? Use Euler's formula to explain.

TASK 6: Analyze with the formula for the Euler Load.

Using Euler's formula, holding E and L constant, the effect of doubling I for a column gives

$$P(2I) = \frac{\pi^2 E\ (2I)}{L^2} = 2 \cdot \left(\frac{\pi^2 E\ I}{L^2}\right)$$ or 2 times the Euler Load of the original column.

What is the effect of tripling the value of I?

What is the effect of halving I?

What if I was decreased by 40%?

In mathematics this type of relationship is called a **linear** relationship.

CONCLUSION

When an architect or engineer designs or chooses a steel column, the choice is based on the load that it must carry. Steel columns are produced in various shapes and lengths. Does the Moment of Inertia or the Effective Length of a column have more effect on the load that a column can carry?

If a column must be redesigned to carry twice the load, how could the length be changed? How could the cross section be changed?

MATHEMATICS LABORATORY INVESTIGATION

BUILDING SITE EXCAVATION

Topic: **MODELING, with SLOPE, INTERPOLATION, VOLUME, 3D GRAPHS**

Prerequisite knowledge: *Geometry, Spreadsheet Fundamentals*

I. INTRODUCTION:

Site excavation is part of the process in new housing construction, building Boston's third harbor tunnel, and environmental reconstruction such as toxic waste clean-up and wetland reclamation. Of course, there are many more projects that require earth to be moved. Here we look at some of the mathematical aspects of site excavation for a new house. We will be using a process called "mathematical modeling." This is a problem-solving strategy that involves solving a simplified version, or "representation," of the problem in order to efficiently obtain useful results without great expenditure of time, money, or effort.

II. THE MODELS

Different kinds of models are often used to study a real world situation before committing equipment and personnel to a task. Often the modeling will take place before a company bids on a project. Several models may be constructed prior to and during a project. The models help the company estimate its costs and make a realistic bid; if a bid is too low the company may win the contract but lose money on the job; if too high, it may not win the contract award. Throughout the computational aspects of what follows runs the thread of modeling a situation in multiple ways.

Sometimes a scale model of a lot is built. It is possible to create other physical models as well as numerical and graphical ones. Some of the possibilities are developed below. The graphical model (or contour map) on the next page shows a diagram of the house lot and the placement of the house's foundation. The contour lines on the diagram show the lay of the land, as determined from a land survey. Along each contour line, the height of the land is the same. Each contour line is labeled with its height in feet above sea level.

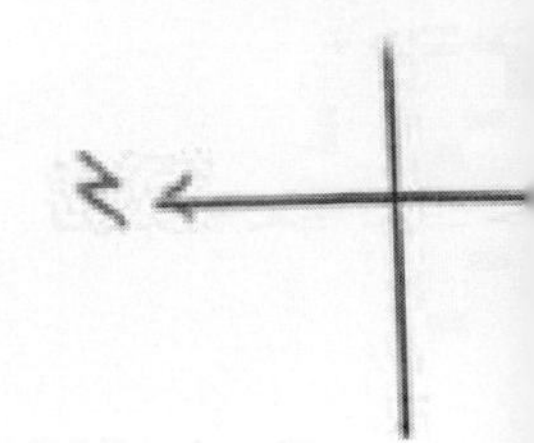
N

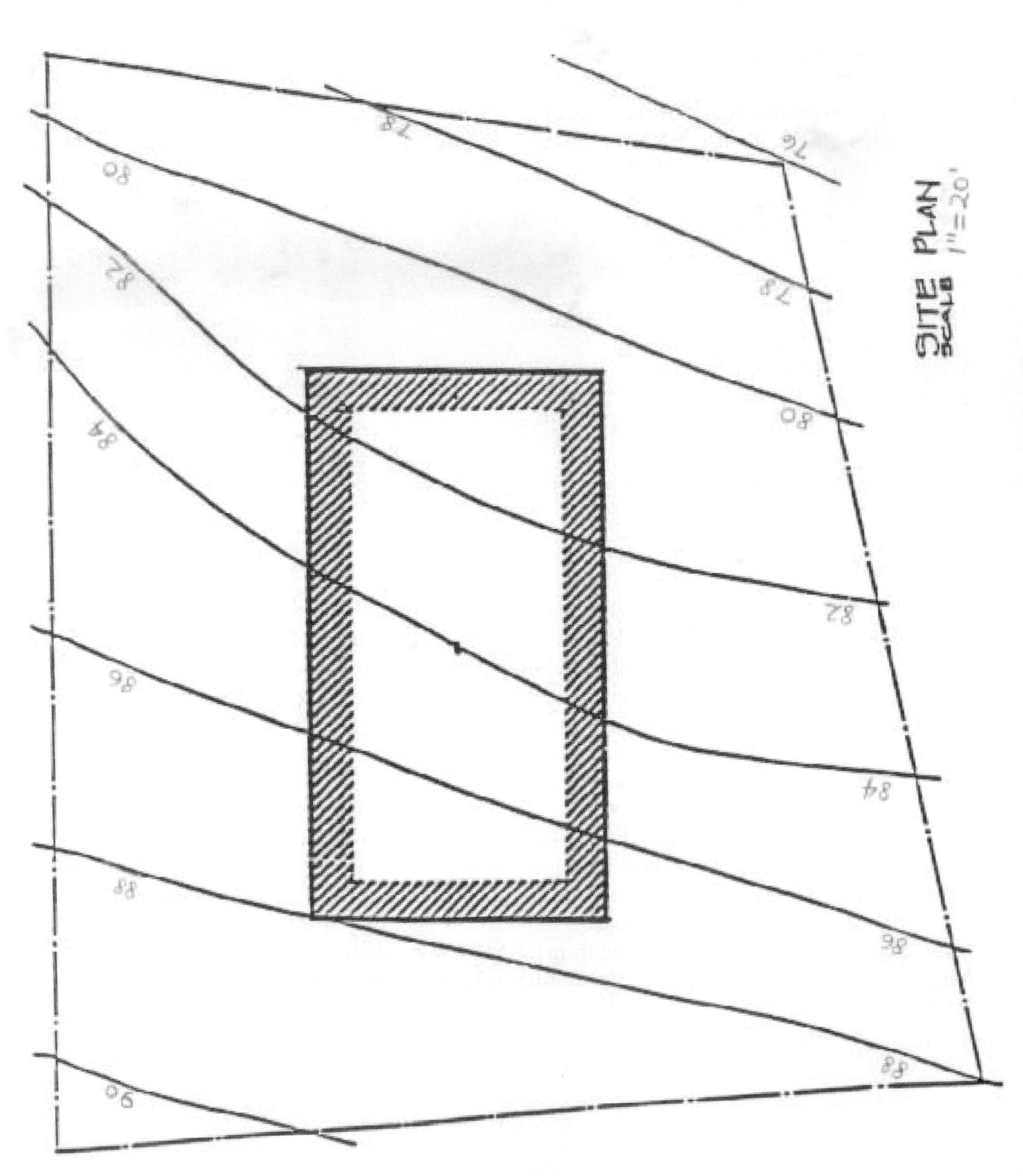
SITE PLAN
SCALE 1"=20'
76
78
78
80
80
82
82
84
84
86
86
88
88
90

III. BUILDING A PHYSICAL MODEL **(To be done <u>before</u> the laboratory session)**

Construct a three-dimensional physical model of the house lot, using cardboard, or some similar material, to build up layers at the various heights indicated on the contour map. That is, for each contour level, use a copy of the site plan as a pattern to cut out a surface. Stack the surfaces for each contour level to create a "step" model of the site.

Compare your model to a physical scale model of the site like the one shown here. Discuss with your teammates what information the physical (cutout) and graphical (contour map) models preserve and what information each model lacks compared to the actual site itself.

Briefly describe your conclusions here:

IV. WORKING WITH GRAPHICAL AND NUMERICAL MODELS

TASK 1: Estimating from a contour diagram.

Take a few minutes to get a general feel for the situation. Don't use the grid, rulers or any other tools to make your estimations.

On the house lot diagram, draw an arrow at the location of steepest slope of the land and also at the location of gentlest slope.

At its steepest, the land falls about _______ vertical feet in about _____ horizontal feet.

Make an initial estimate of the elevation of the land at the four corners of the house's foundation.

NW corner _____feet NE corner _____feet

SW corner _____feet SE corner _____feet

What is the approximate average elevation of the land where the foundation is to be laid?

_____feet

Describe briefly how you arrived at this estimate.

TASK 2: Estimating distances using a grid.

Examine the transparent grid, which is constructed to the same scale as the house's site plan (1 inch equals 20 feet).

Use the grid to estimate the lengths of the boundaries of the property.

southern side length = _____ feet northern side length = ______ feet

eastern side length = ______ feet western side length = ______ feet

Use the grid to estimate the dimensions of the foundation.

length = _____feet, width = _____feet.

Now see if you can use the grid to check/refine the estimates of the land elevation at the four corners of the foundation.

NW corner _____feet NE corner _____feet

SW corner _____feet SE corner _____feet

Use the grid to refine your estimate of the average elevation of the land where the foundation is to be laid.

_____feet

Describe the method you used to find this average elevation.

Now use the grid to check or refine your previous estimate of the slope of the land at the site.

At its steepest, the land falls about _______ vertical feet in about _____ horizontal feet.

Form the ratio of fall to run_____. In technical terminology, this ratio is called ***slope***, indicated either as (*rise/run*) or (*fall/run*).

The following is a sketch of the cross-section of the land along the north edge of the foundation. The land falls about 6.6 vertical feet in 75 horizontal feet. It has an average slope of $\frac{-6.6\,ft}{75\,ft} = -0.088$.

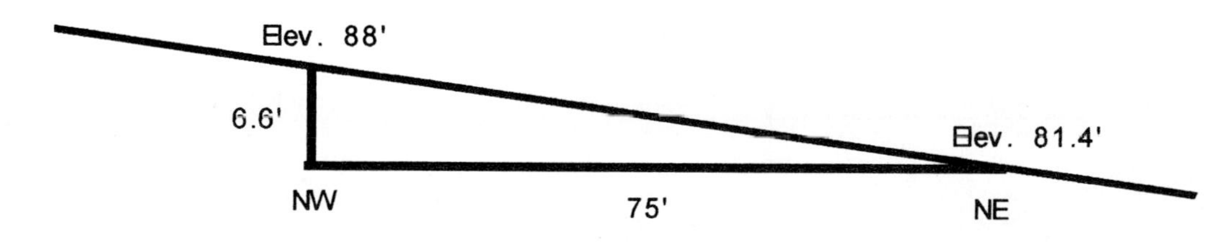

Sketch a similar diagram for each of the other three edges, and determine the average slope of the land along each edge.

slope on south foundation edge: ______

slope on west edge: _______

slope on east edge: ________

TASK 3: Estimating area and volume

Using your refined estimates from TASK 2, estimate the area of land covered by the foundation: ________square feet.

If the bottom of the foundation is to be laid at 84 feet above sea level, the contour diagram shows some dirt will have to be excavated beneath one portion and some fill will be needed beneath another portion of the foundation.

Using the contour diagram and the grid, does it look like more dirt will be excavated than is needed to support the other section of the foundation? **YES** or **NO**? **Explain your answer.**

Use the refined estimated elevation of the land (from TASK 2) at the foundation location to determine if fill will have to be <u>brought to</u> or <u>taken from</u> the foundation site if the base of the foundation is at 84 feet above sea level.
Show your work.

CONCLUSION: **Fill Needed _____** **Excess Fill _____**
(check one)

Estimate the volume of dirt that will have to be brought in or taken away, and **explain your method.**

________ cubic feet

TASK 4: <u>Further refining the estimates.</u>

Overlay the transparent grid on the site plan in such a way that the origin (0,0) lies at the southwest corner of the foundation, and the x axis lies along the southern edge of the foundation.

Along the western end of the south wall, the land falls 2 vertical feet over 15 horizontal feet, but note that the slope is <u>not</u> constant over the entire length of the wall. Proportionality can be used to estimate the height of the land at the other points along the south wall, as shown in the diagram below. If we assume a constant slope between each pair of contour lines, then similar triangles can be formed, as in the diagram, to compute the heights at other points between the contours.

For example, with our grid placed as above, the 82' contour line crosses the x axis at x=52. This corresponds to a point in three dimensions, $(x, y, z) = (52, 0, 82)$ where z represents the elevation in feet. Similarly, the 80' contour line crosses the x axis at $(82, 0, 80)$. Note that this is past the southeast corner of the foundation, which lies at approximately $x = 76$. This can be used to estimate the elevation of the southeast corner.

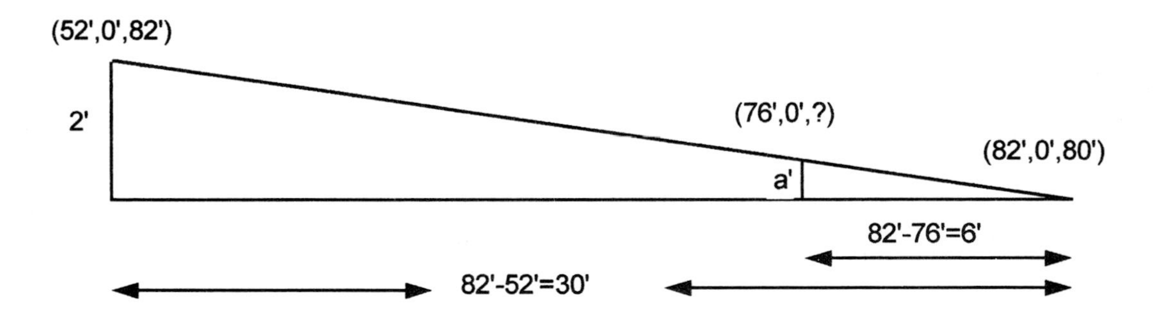

Notice that the elevation drops by 2 feet over a 30 foot run which can be used to infer how much it would drop over a 6 foot run.

$$\frac{a}{6} = \frac{2}{30} \quad \text{so} \quad a = \frac{12}{30} = 0.4.$$

Therefore the elevation is 80+0.4 or 80.4'.

Verify that the elevation at $x = 70$ along the south foundation wall will be 80.8'
Show your work below.

Repeat this process to estimate the elevation of every grid point along the south wall:

x	0	10	20	30	40	50	60	70	80
y	0	0	0	0	0	0	0	0	0
z								80.8	

This process of using proportionality to find intermediate values is called *linear interpolation.*

Now assume a constant slope (fall/run or rise/run ratio) between each pair of contour lines, and compute heights at all grid points in the foundation by interpolation. **Show your results here, and discuss them with your group to verify your method.**

	x=0	10	20	30	40	50	60	70	80
y=40									
30									
20									
10									
0									

Table of Elevations at Grid Points of the Foundation (in feet)

Your data could be copied into a spreadsheet. Shown below are some different types of graphs that can be generated from that data:

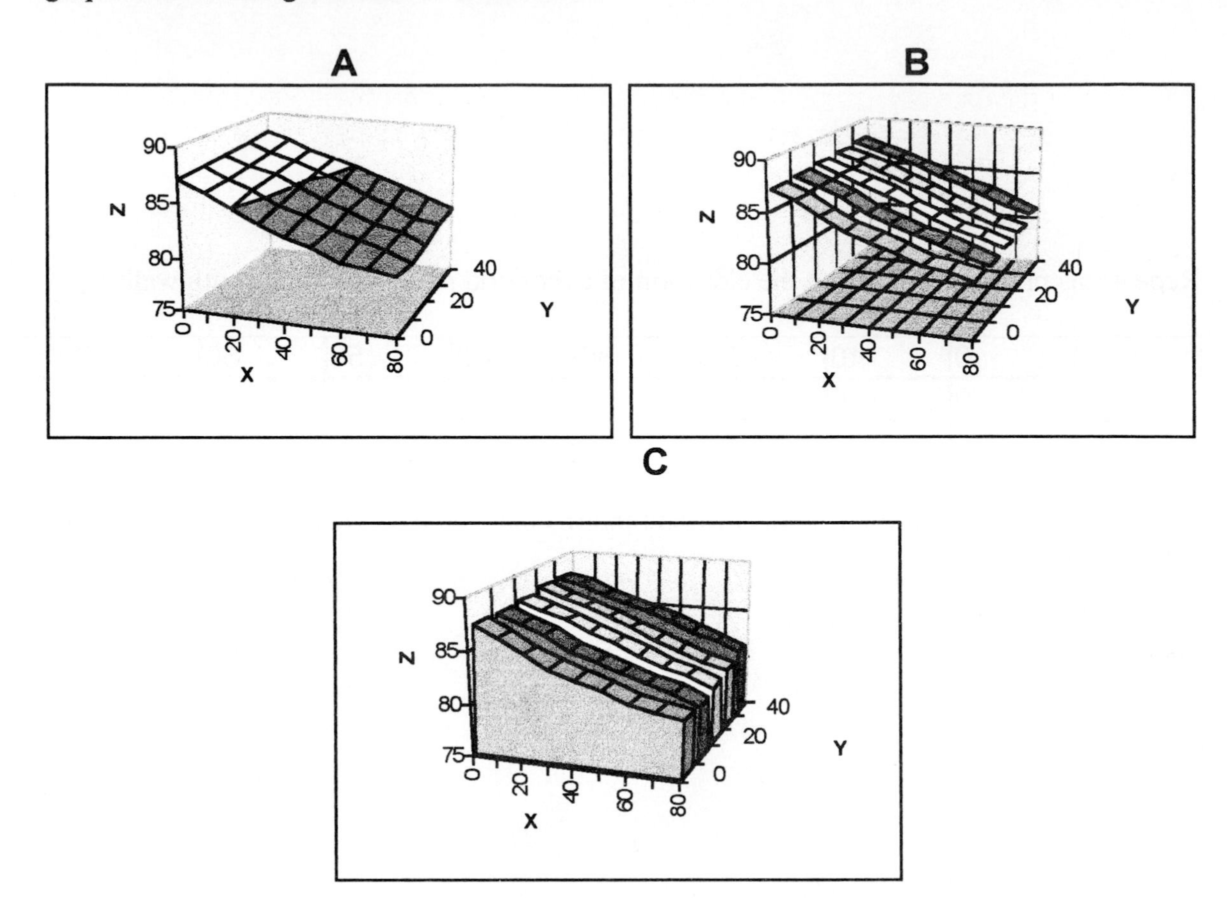

You have now seen several types of models: these graphs coming from a numerical model, the contour map, and the physical model. **Discuss which type (or types) of model does the best job of picturing the lay of the land, and why.**

Now, make <u>your own</u> three-dimensional graph of your spreadsheet data, and include it with your report.

Finally, use the spreadsheet's computational ability to determine the average height of each column of earth, measured from the base of the foundation. (Some of the columns will have "negative" heights.) Using the "base area" of each column, calculate a final estimate of excess or needed fill volume. Include a printout of any spreadsheets used in your report.

V. ADDITIONAL ACTIVITIES

A. Common soil will typically swell in volume by 25% after it is removed from the ground, and loose soil can be compacted 10% when used as fill. How will this affect your conclusions?

B. Make allowances for <u>banking</u> of the earth along the edges of the foundation.

C. <u>Cost considerations:</u> Assume that the market value of fill is $27.35 per cubic yard, if you need to purchase extra fill. However, if your excavation generates excess fill you can sell it at 80% of the market value, but it will cost you $14.80 (per cubic yard) in labor and other costs to sell the excess. Determine your potential "fill-related" cost or net profit.

D. Discuss the errors and assumptions inherent in the investigation, as well as the accuracy of the calculation as revealed by comparison of your results with those of other groups.

MATHEMATICS LABORATORY INVESTIGATION

COMPUTING INTEREST – I

Topic: **THE CONVOLUTION SUMMATION, DISCRETE VARIABLES**

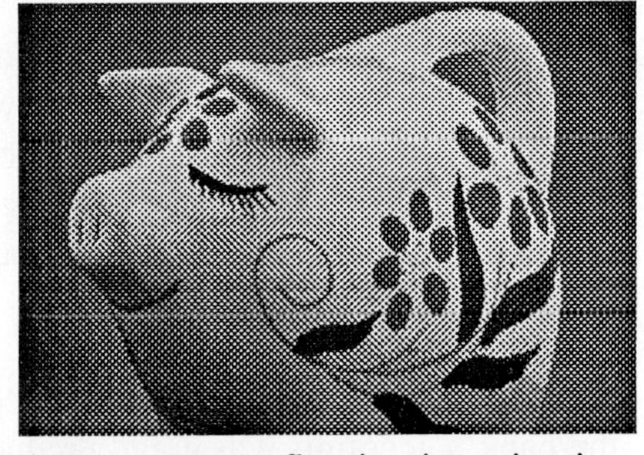

Prerequisite knowledge: *Percentage calculations, Graphing functions, Shifting and Reflecting graphs, and Functional Notation.*

Equipment: Graph paper, a transparent grid and marker, and a calculator.

I. INTRODUCTION

Several systems such as a mass oscillating on a spring, or current flowing in a circuit that includes resistors, inductors, and capacitors, or fluid flowing through a network of pipes are of great interest to engineers and engineering technologists and are the subject of much study. Each of these engineering systems is considered to have an input and an output. For the mechanical system, the input may be a force suddenly applied to the mass and the output to be analyzed could be the resulting velocity of the mass. For the electrical circuit, the input could be a voltage source, E(t), which provides energy to an electrical network while the output might be the voltage drop, $V_r(t)$, across a resistor in the circuit. For the hydraulic example, the input could be the abrupt closure of a flow valve at some point in the system and the output could be the resulting shock wave which would be transmitted throughout.

The three engineering systems mentioned above possess a memory component. The output of an engineering system which possesses a memory component is the result of the present input to the system as well as the residual effect of the previous inputs. In other words, the history of <u>all</u> past inputs to the system contributes to the total output of the system at any given point in time.

A very sophisticated mathematical construct called the CONVOLUTION INTEGRAL has been developed to keep track of the residual effects of previous inputs to an engineering system as well as the effect of the present input to the system. The CONVOLUTION INTEGRAL deals with continuous variables and is developed within the context of a Calculus course. Since you have not yet reached that level in your Mathematics education, this Laboratory Investigation will introduce you to a similar mathematical device that deals with discrete variables and is called the CONVOLUTION SUMMATION.

Your understanding of the CONVOLUTION SUMMATION will not only provide you with a powerful tool which you are now ready to use but will also prepare you for the study and full appreciation of the CONVOLUTION INTEGRAL which you might study in a year or two.

II. THE HISTORY OF COMPUTING INTEREST

Approximately 40 years ago computers were not yet available to be used in businesses such as banks. Interest calculations were performed with the use of adding machines and/or pencil and paper. Computing the amount of interest earned in all of the savings accounts in a bank was a slow and time consuming process. Consequently, interest was calculated and credited to a savings account only four times per year. We describe this as compounding quarterly.

As computers were introduced into banks, the time required to compute interest for the bank's accounts was greatly reduced. Interest could now be calculated and credited to savings accounts more frequently - monthly. Accordingly, interest was compounded monthly. High-speed computers have now replaced the slower original computers and interest is now computed and credited to your account daily (compounded daily) although it is still posted to the account only on a monthly basis.

In this Lab Investigation the main goal is to familiarize you with the CONVOLUTION SUMMATION. Consequently, we are going to **assume** that the interest is compounded only once each month to simplify **your** computations.

III. THE PROBLEM

You will determine the amount of interest credited to a savings account in a typical month.

Deposits are going to be made into a savings account that pays interest at a 5% yearly rate. The bank calculates interest on the 24th day of the month. However, if the 24th day of the month falls on a non-business day (Sundays or holidays), interest is calculated on the first business day after the 24th. Interest is calculated based on the number of days that a deposit has been in the account. A $100 deposit on account for 10 days would earn twice as much interest as a $100 deposit on account for 5 days., etc.

The following is a deposit history for one monthly period.

Deposit Date	Deposit Amount
2/27/96	$150.00
3/07/96	$350.00
3/15/96	$250.00
3/20/96	$425.00
3/23/96	$200.00

FEBRUARY 1996

SUN	MON	TUE	WED	THU	FRI	SAT
				1	2	3
4	5	6	7	8	9	10
11	12	13	14	15	16	17
18	19	20	21	22	23	24
25	26	27	28	29		

MARCH 1996

SUN	MON	TUE	WED	THU	FRI	SAT
					1	2
3	4	5	6	7	8	9
10	11	12	13	14	15	16
17	18	19	20	21	22	23
24	25	26	27	28	29	30
31						

IV. LET'S TALK IT UP

Get together with your lab partners at the beginning of the lab or preferably before the lab period and discuss the problem. Try to arrive at a solution. Some of you may have different approaches . Try more than one. See where you agree and where you disagree.

How much interest (in dollars and cents) did you figure to be credited to the account?

Do you think that your solution is correct to the nearest cent?

Briefly describe the procedure that you used to arrive at your solution.

What assumptions did you make in arriving at your solution?

Be prepared to make a short presentation to the class.

During the remainder of this Laboratory Investigation, you will try to find a solution to the problem by means of both a graphical approach and a symbolic approach.

V. A GRAPHICAL APPROACH TO THE PROBLEM.

TASK 1: Preparing Graphs.

On graph paper create a graph for the **Deposit Function**. Be sure to carefully label and increment the horizontal and vertical axes. The independent variable should represent the number of days into the interest cycle (Feb. 26 is day 2 and Feb 29 is day 5). You may want to consider the questions following the next paragraph while preparing the graph.

Create a graph on a transparent grid that represents the amount of interest earned on a $1.00 deposit into the savings account . The independent variable should indicate the number of days that the $1.00 deposit has been in the account. In order to facilitate calculations, the interest earned should be kept in fractional form, not converted to a decimal. Be sure to carefully label and increment the horizontal and vertical axes. Let's call this function the **Unit Interest Function**.

What minimum and maximum values did you choose for the horizontal axes of the Deposit Function and the Unit Interest Function?

Defend your choice of these minimum and maximum values by presenting scenarios in which these values would actually occur.

Consider the variable depicted on the horizontal axis of the Unit Interest Function. How many different values may this variable assume?

Variables are often referred to as being either "continuous" or "discrete". Would you say that the variable depicted along the horizontal axis of the Unit Interest Function is a continuous variable or a discrete variable. Defend your answer.

TASK 2: <u>Manipulating Graphs.</u>

Now, as a group, try manipulating the graphs of the Deposit Function and the Unit Interest Function in order to find a solution to the problem. Write a verbal description of the manipulations and be prepared to make a presentation to the class.

Explain why the graph manipulations yield a method of performing the necessary computations for the desired solution.

Do you think your solution is correct to the nearest cent? If not, make the necessary changes to get a solution correct to the nearest cent and describe what changes you had to make and why.

Compare your solution with your original estimate of the solution and explain any differences.

VI. A SYMBOLIC APPROACH TO THE PROBLEM.

TASK 3 : Symbolically Representing the Unit Interest Function.

Using function notation, write the equation of the Unit Interest Function.

Describe the domain of the Unit Interest Function.

TASK 4 : Symbolically Representing the Shifted and Reflected Unit Interest Function

(Note: The order of the following two steps may be reversed.)

Using function notation, write the equation of the Unit Interest Function after it has been slid (shifted) the desired number of days to the right.

Using function notation, now write the equation of the Unit Interest Function after it has been flipped (reflected) through the desired vertical line.

TASK 5 : Symbolically Representing the Solution.

Using the function notation for the Deposit Function and the Unit Interest Function that has been shifted and reflected, symbolically represent the solution to the problem. (Hint: the representation is called a sum of products.)

TASK 6 : Arriving at the Convolution Summation.

The solution to our problem is a sum of five products which correlates to the five days on which deposits were actually made. Of course if deposits were made on ten days, then our solution would be a sum of ten products.

In order to allow for a more general situation, let's say that on a day for which there is no deposit, we enter the deposit for that day as $0.00. Then on a thirty day interest cycle the solution to the problem would be the sum of thirty-one products, remember day 0, many of which might be $0.00 .

Symbolically rewrite the solution to the problem as a sum of products, including the products for days for which no deposit was made.

The sum of products which you have just wriitten is called a **Convolution Summation**. Note that each day in the interest cycle is represented by a product in the sum. There are as many products in the sum as there are days in the interest cycle. Actually, there is an extra product as we have allowed for day 0. When the actual computation is performed, all but five of the products will be $0.00 .

You should also note that the **Convolution Summation** can also be seen graphically when you shift and reflect the Unit Interest Function and lay it on the Deposit Function. Again, counting no deposit days as $0.00 deposit days, you can visualize products on every day of the interest cycle. The sum of these products yields the desired result.

VII. WRAP-UP

In the COMPUTING INTEREST II Laboratory Investigation you will be shown how to express the Convolution Summation in short-hand form by using a mathematical symbol called the **SUMMATION SIGN**. You will also be shown how to handle withdrawals from the savings account. And finally you will be shown a spreadsheet approach to the problem.

MATHEMATICS LABORATORY INVESTIGATION

COMPUTING INTEREST - II

Topic: **THE CONVOLUTION SUMMATION** continued, introduction to the summation symbol, the commutative property of convolution, and use of a calculator or computer spreadsheet as a solution tool.

Prerequisite knowledge: *Completion of the preceding COMPUTING INTEREST - I Investigation. Familiarity with a graphics calculator or computer spreadsheet application.*

Equipment: Graphics calculator or computer spreadsheet application

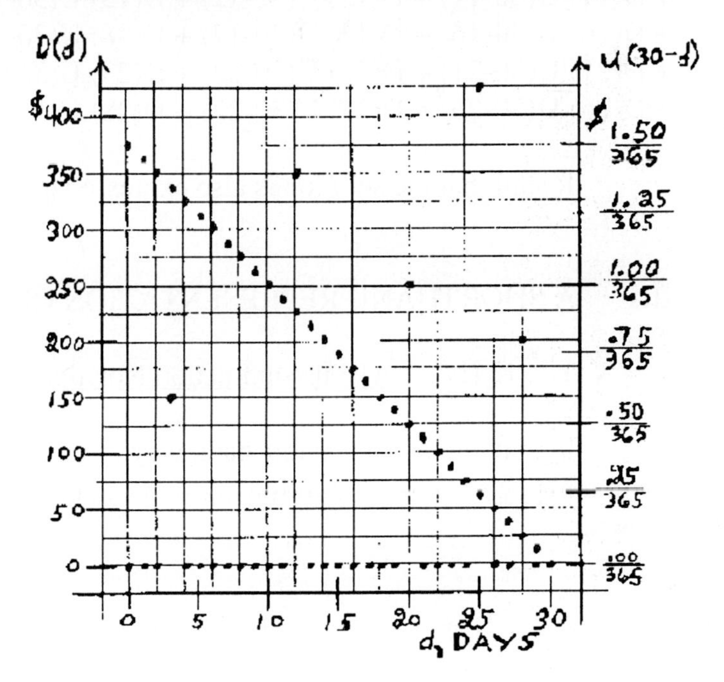

I. INTRODUCTION

This lab is an extension of the Computing Interest - I Investigation. It introduces some of the mathematical symbols used in expressing convolution and also explores the principle of commutativity.

In addition to regular deposits, banking activity would also include withdrawals. This lab requires you to develop a systematic and unified approach to handle deposits, withdrawals and no activity at all on a daily basis. In an actual banking environment, interest is compounded and computed daily. This is possible as a result of the high speed computers which are available in today's technology. Sophisticated formulae are programmed into computers to perform these calculations, As in the Computing Interest -I Investigation, for learning purposes only, we will assume that interest is calculated and credited to the account only one day per month. This is called compounding monthly.

Finally, this lab demonstrates how a spreadsheet can be used to visualize the mechanics of convolution.

II. A REVIEW OF COMPUTING INTEREST - I INVESTIGATION RESULTS

Recall that the solution to the problem posed in the Computing Interest - I Investigation is expressed symbolically as follows, where D represents the Deposit Function and U represents the Unit Interest Function.

Interest =
D(0) U(30-0) + D(1) U(30-1) + D(2) U(30-2) + D(3) U(30-3) + D(4) U(30-4) + D(5) U(30-5)
+ D(6) U(30-6) + D(7) U(30-7) + D(8) U(30-8) + D(9) U(30-9) + D(10) U(30-10)
+ D(11) U(30-11) + D(12) U(30-12) + D(13) U(30-13) + D(14) U(30-14) + D(15) U(30-15)
+ D(16) U(30-16) + D(17) U(30-17) + D(18) U(30-18) + D(19) U(30-19) + D(20) U(30-20)
+ D(21) U(30-21) + D(22) U(30-22) + D(23) U(30-23) + D(24) U(30-24) + D(25) U(30-25)
+ D(26) U(30-26) + D(27) U(30-27) + D(28) U(30-28) + D(29) U(30-29) + D(30) U(30-30).

Recall that the solution is also expressed graphically on page 1 of this investigation.

III. A SHORTHAND REPRESENTATION OF CONVOLUTION

TASK 1: Using Summation Notation.

In mathematics the short-hand way of representing a long sum is with the use of the summation symbol Σ. The following sum can be abbreviated by using the summation symbol as shown:

$$0\times(-1) + 1\times0 + 4\times1 + 9\times2 + 16\times3 + 25\times4 + 36\times5 = \sum_{n=0}^{6} n^2(n-1)$$

The right hand side is read as "the sum from n equals 0 to 6 of the function $n^2(n-1)$". This means that n takes on all of the integer values starting at 0 and ending at 6, inclusive. The function is evaluated for each of the individual values of n and then each of these results is summed to obtain the final result which happens to be 350 in this example.

You can now write the solution to the interest problem symbolically using summation notation. Go ahead and do it below. You will probably use a variable other than n.

TASK 2: Using Convolution Notation.

The summation which you have derived is called the **convolution summation**. Mathematically, given two functions f(t) and g(t), the convolution summation of f(t) and g(t), calculated at t = 30, is defined and denoted as follows:

$$(f * g)(30) = \sum_{t=0}^{30} f(t) \cdot g(30 - t),$$

where * is the symbol for convolution and is read as "f convolution g".

Symbolically express the solution to the problem as the convolution of the Deposit Function and the Unit Interest Function.

IV. COMMUTATIVITY OF CONVOLUTION SUMMATION

TASK 3: Determining whether the Convolution Summation is commutative.

Instead of shifting and reflecting the Unit Interest Function to set up the products try placing the Deposit Function over the Unit Interest Function and shifting and reflecting the Deposit Function. What result do you get?

Is it the same or different than our previous result? Why?

Symbolically express this solution as a convolution of the Unit Interest Function and the Deposit Function.

Is $(D * U)(30) = (U * D)(30)$? In other words, is the convolution summation commutative? Defend your conclusion.

III. A SPREADSHEET APPROACH TO THE PROBLEM

TASK 4: A spreadsheet solution to the problem.

Using your calculator or a computer software program create a spreadsheet that illustrates the solution to this problem. Even though there are only five deposits during the interest cycle, set up your spreadsheet to allow for deposits on any given day in the interest cycle.

Show the **outline** of your spreadsheet to include how the column values are determined.

IV. ALLOWING FOR WITHDRAWALS

TASK 5: Withdrawing Money From the Account.

Suppose the following withdrawals were made from the account.

Date of withdrawal	Amount of withdrawal
3/04/96	$ 25.00
3/12/96	$ 25.00
3/19/96	$ 35.00
3/21/96	$ 45.00

Determine the amount of interest that is posted to the account in March.

Explain how you treated the withdrawals in your solution.

How would you handle deposits and withdrawals made on the same day?

V. A POSSIBLE EXTENSION

You might try writing a computer program to arrive at a solution of this problem.

VI. HOMEWORK

TASK 6 : Reporting the Solution.

Create a hard copy of the spreadsheet which represents the solution to this problem.

TASK 7 : Summarizing your understanding.

Explain your understanding of the convolution summation. Try to determine a specific application for which the convolution summation would be useful.

MATHEMATICS LABORATORY INVESTIGATION

CONCRETE STRENGTH TESTING

Topic: **INTERPRETATION OF STANDARD DEVIATION, FRACTIONAL EQUATIONS**

Prerequisite knowledge: *Mean & Standard Deviation Calculations, Histograms*

I. INTRODUCTION

The purpose of strength testing is to ensure that concrete placed in the field is in compliance with the engineer's design. Variations can occur in the strength of concrete due to changes in the weather, ingredient proportions, mixing, transporting, placing and curing. One way of verifying the strength of the concrete used is to take samples of concrete at the site and then break these test samples in a laboratory. Technicians measure and record the compressive strength of each sample. By taking many samples and through the use of statistics engineers are able to determine the variability of the produced concrete. They can then predict the strength of the concrete placed in the field.

The following is a quote from the American Concrete Institute's Concrete Laboratory Testing Technician Workbook:

> "Statistical procedures provide tools of considerable value in evaluating results of strength tests and information derived from such procedures is also of value in refining design criteria and specifications."

II. SAMPLE GATHERING AND LABORATORY TESTING

Concrete samples are gathered in the field at the time of the concrete placement directly from the concrete truck. It is the job of the concrete inspector to properly label and store the concrete samples in the same manner as the actual deposited concrete. These samples (6" diameter cylinders - 12" in height) are delivered to the laboratory where they are broken, usually at 28 days (the standard time required to properly cure concrete). Samples are taken daily and in the case of large concrete placements may be taken more frequently. Enough samples need to be taken to insure an accurate representation of the variations within the concrete. Concrete samples are not only used for measuring the concrete design, but also for determining when it is safe to remove the concrete forms.

The properly stored concrete cylinders are broken at the designated time in a properly calibrated testing machine. The results of these breaks are the data that are used for measuring the variation of the supplied concrete as well as its strength.

III. STATISTICAL CALCULATIONS

The data below was collected from the Boston Central Artery / Tunnel Project. These numbers represent the results of cylinder break strengths due to compressive loading on N = 50 cement cylinders from one concrete mix.

5057 psi
5089 psi
5297 psi
5779 psi
4832 psi
5584 psi
5151 psi
4674 psi
4393 psi
4565 psi
4158 psi
5569 psi
5168 psi
5131 psi
5312 psi
4839 psi
5187 psi
4500 psi
5111 psi
4758 psi
4420 psi
5403 psi
5526 psi
4732 psi
4615 psi

4676 psi
4776 psi
5252 psi
5274 psi
6017 psi
5056 psi
4817 psi
4697 psi
4004 psi
4854 psi
4985 psi
4579 psi
4259 psi
4875 psi
5518 psi
4691 psi
5760 psi
4373 psi
5049 psi
4159 psi
4706 psi
5332 psi
4329 psi
4065 psi
4969 psi

Task 1: Examining the Concrete Strength Data

The average strength (*the mean*) of all individual tests is:

$$\overline{X} = \frac{x_1 + x_2 + \cdots + x_N}{N} = \frac{1}{N}\sum_{i=1}^{N} x_i$$

Compute the average strength of the test results above. Write a few sentences explaining the significance of this value.

Draw a histogram of the data.

The measure of dispersion of the data is the root mean square of the deviation of the strengths from their average. This is known as the standard deviation. There are two formulas for calculating the standard deviation; two formulas for one quantity can be confusing. One formula is for what is called the *population standard deviation* and is usually denoted by the Greek letter sigma, σ or σ_N.

$$\sigma = \sqrt{\frac{\sum_{i=1}^{N}(x_i - \overline{X})^2}{N}}$$

The other formula is for the *sample standard deviation* and is generally denoted in mathematics by the letter S or S_{N-1}.

$$S = \sqrt{\frac{\sum_{i=1}^{N}(x_i - \overline{X})^2}{N-1}}$$

Notice the difference in the formulas is only the N or N-1 in the denominators. If N is large enough, both formulas give essentially the same results.

Calculate both the *population standard deviation* and the *sample standard deviation* for the concrete strength test data and compare results.

σ = __________ and S = __________.

$|\sigma - S| =$ __________.

It is common to see only the *sample standard deviation* used in engineering applications. Because of this, engineers often use the symbol σ or σ_{N-1} for the *sample standard deviation* which mathematicians denote by S. This laboratory investigation will look at the difference between both formulas but always use the *sample standard deviation* and denote it by S in analyzing data. In technical applications be sure you know which standard deviation formula is intended when the symbol σ is used.

Before proceeding to analyze the concrete strength data, consider a data set with very different characteristics to see how the statistics differ. The following data taken from the Sports Illustrated 1994 Multimedia Sports Almanac are winning times for the men's 100 meter run in Summer Olympic Games from 1896 through 1992.

Year	Winner	Time(sec.)
1896	Thomas Burke, United States	12.00
1900	Frank Jarvis, United States	11.00
1904	Archie Hahn, United States	11.00
1906	Archie Hahn, United States	11.20
1908	Reginald Walker, South Africa	10.80
1912	Ralph Craig, United States	10.80
1920	Charles Paddock, United States	10.80
1924	Harold Abrahams, Great Britain	10.60
1928	Percy Williams, Canada	10.80

1932	Eddie Tolan, United States	10.30
1936	Jesse Owens, United States	10.30
1948	Harrison Dillard, United States	10.30
1952	Lindy Remigino, United States	10.40
1956	Bobby Morrow, United States	10.50
1960	Armin Hary, West Germany	10.20
1964	Bob Hayes, United States	10.00
1968	Jim Hines, United States	9.95
1972	Valery Borzov, USSR	10.14
1976	Hasely Crawford, Trinidad	10.06
1980	Allan Wells, Great Britain	10.25
1984	Carl Lewis, United States	9.99
1988	Carl Lewis, United States	9.92
1992	Linford Christie, Great Britain	9.96

Task 2: Examining the Winning Time Data

Compute the average winning time of the results above. Write a few sentences explaining the significance of this value and what this average does not tell you about future winning times.

Draw a histogram of the winning time data. Write a few sentences that explain why the histogram of this data has a very different shape from the concrete strength data histogram.

Calculate both the *population standard deviation* and the *sample standard deviation* for the winning run time data and compare results.

σ = __________ and S = __________.

$|\sigma - S| =$ __________.

Notice the absolute difference between σ and S is much larger for the concrete strength data than for the winning run time data. With more data samples (N=50) the concrete strength *population* and *sample standard deviations* should be closer than for the winning run times where N = 23. To see this is, normalize the differences by their respective means. That is, divide by the means so the typical (mean) data values are "factored out" of the standard deviation values.

Task 3: Comparison of *Population* and *Sample Standard Deviations.*

Calculate the normalized differences between population and sample standard deviations

for the concrete strength data $\frac{|\sigma - S|}{\overline{X}} =$ __________

and for winning run time data $\frac{|\sigma - S|}{\overline{X}} =$ __________.

Convert the above results to percentages of the means:

for concrete strength:

for winning run times:

Write a few sentences explaining why normalizing lets you compare results for the different data sets. Also explain why one percentage is higher than the other.

Task 4: Concrete Strength Variation

The standard deviation expressed as a percentage of the mean is called the coefficient of variation. Notice, to compare differences in *population* and *sample standard deviations* you calculated percentages of the mean of each data set. The coefficient of variation in strength of concrete within a single test data set is found by the same normalization process. The coefficient of variation is calculated as follows:

$$V = \frac{S}{\overline{X}} \cdot 100$$

Compute the coefficient of variation for the concrete strength test results.

Suppose the coefficient of variation for concrete strength tests for a different construction project was calculated to be 8.3. Explain why you think there would be more or less variability in the concrete strength test data for this second construction project.

Explain why normalizing the standard deviation to the mean lets you compare variability results from different projects.

The range of variation is found by subtracting the lowest value of test data from the highest value. What is the range of variation for the concrete strength test data?

Since you are concerned with only one set of test data to determine the strength of concrete used in this construction project, the standard deviation is the most commonly used measure of variation from the mean. The value of the standard deviation measures how closely the values of the data set are clustered about the mean. A small value means the data set are clustered more closely about the mean, and a large value means the data are more scattered. The design engineer's specifications for strength of concrete used in the project will require that a certain percentage of the test samples lie within a certain number of standard deviations of the mean.

By using the mean and the standard deviation, the proportion or percentage of the data set that fall within a given interval about the mean may be calculated. **Chebyshev's Theorem**, named after the Russian mathematician P. L. Chebyshev (1821 - 1894), gives a rough estimate of the fraction of values that fall within k standard deviations of the mean. K must be a value larger than 1.

CHEBYSHEV'S THEOREM

For any number k greater than 1, a fraction equal to at least $1 - \frac{1}{k^2}$ of the data values lie within k standard deviations of the mean.

Example: Let k = 2 and apply Chebyshev's theorem.

$$1 - \frac{1}{k^2} = 1 - \frac{1}{(2)^2} = 1 - \frac{1}{4} = \frac{3}{4} = 0.75 \text{ or } 75\%$$

Therefore at least 75% of the data points in any data set must be within 2 standard deviations of the mean.

Task 5: Practice Using Chebyshev's Theorem on Concrete Strength Data.

Determine the range of values between $\overline{X} \pm 2S$.

Look at the data. How many actual values are in this range?

Does Chebyshev's Theorem hold? Explain.

What is the range of values between $\overline{X} \pm 2.5S$?

Look at the data. How many actual values are in this range?

Does Chebyshev's Theorem hold? Explain.

If a sample contains the middle 89% of the data, how many standard deviations does Chebychev's Theorem say the data must lie within? (Find k)

If a sample contains the middle 95% of the data. How many standard deviations on each side of the mean is this? (Find k)

Task 6: Required Average Strength of Concrete

The strength of the control test cylinders is the only tangible evidence of the quality of the concrete used on a project. The number of tests lower than the desired strength is more important in computing the load bearing capacity of a structure than the average strength obtained. Factors of safety allow for some acceptance of singular tests below a specified strength. In many cases test errors may occur, the concrete used may be in a non-critical area, and/or concrete in excess of specifications may be used alongside and offset the weaker batch. The calculation of the required average concrete strength is the adjustment used to account for the variations experienced in concrete batches.

The American Concrete Institute building code requires that compressive strength of concrete equal the larger outcome of the following 2 formulas:

$$f'_{cr} = f'_c + 1.34S$$

or

$$f'_{cr} = f'_c + 2.33S - 500$$

f'_{cr} = strength of concrete required

$f_c' =$ strength of concrete specified

Given the data above and a hypothetically specified strength of 4000 psi, what is the required strength of concrete?

Using Chebyshev's Theorem, what can you conclude about the chance of concrete from this mix exceeding a hypothetically specified strength of 3,500 psi ?

EXTENSIONS

A. Determine an algebraic relationship between the two formulas for standard deviation. That is, write a formula which expresses one in terms of the other.

B. Use only the most recent five years of winning run time data to calculate the mean and *sample standard deviation*. Explain the changes in the mean and standard deviation from the full set of winning run time data. Why should these differences be expected? Use only the last 5 entries [in column 2] of the concrete strength data to calculate the mean and standard deviation. Explain why something similar to the winning run time data did not occur in this case.

MATHEMATICS LABORATORY INVESTIGATION

CONSTRUCTION PROJECT COST AND MACHINE TOOLING

Topic: **OPTIMIZATION OF DISCRETE AND CONTINUOUS FUNCTIONS**
Prerequisite knowledge: *Algebra, Graphing*

I. INTRODUCTION:

Every company seeks to organize its projects in the most efficient manner. The best-run projects are those that are planned out prior to the work beginning. These must be scheduled and estimated in advance, with the most cost-efficient work schedule chosen. Efficiency can be defined either in terms of the time it takes to complete the project or in terms of the cost of the project. In this lab, you will be analyzing efficiency both ways using both discrete and continuous functions. As an example of using discrete functions, the construction costs of a bathroom renovation will be examined. For continuous functions, the process of machine tooling metal parts will be considered.

II. OPTIMIZATION OF DISCRETE FUNCTIONS

In this section, the problems of optimizing the schedule of a construction project and optimizing the total cost will be considered. In these problems, the times involved in the project are all measured in terms of whole days (rather than fractional days). Consequently, the functions involved (such as the number of days required to complete the job) will be integer-valued. This is an example of a *discrete function*.

TASK 1: SCHEDULING

When scheduling a construction project, it is helpful to have a way of describing and visualizing the tasks involved. Before proceeding to the bathroom renovation project, look first at the following simpler example involving the installation of a satellite dish.

The project involves several different tasks. They are:

T_1: Writing the specifications for the construction
T_2: Choosing a site for the dish
T_3: Choosing a vendor for the dish
T_4: Preparing the site
T_5: Purchasing the dish
T_6: Installing the dish

Additionally, there are certain prerequisite relationships between these jobs. For instance, the dish cannot be installed until after it is bought. A complete list of the prerequisites for this job might be:

T_1 must be completed before T_2 or T_3 can be started.
T_2 must be completed before T_4 can be started.
T_3 must be completed before T_5 can be started.
T_4 and T_5 must be completed before T_6 can be started.

All of these relationships can be expressed in the following *network diagram*.

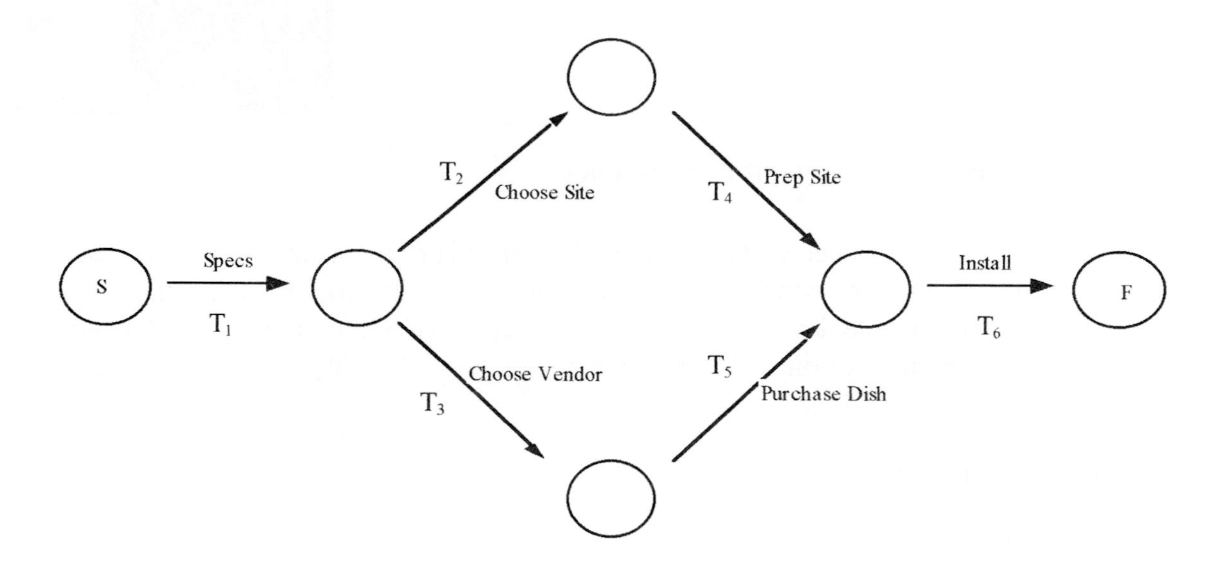

The circles in this diagram (called *nodes*) represent transition times between the various tasks. The node labelled S represents the project's starting time and F represents its finishing time. A node with two arrows (called *edges*) coming out of it, such as the one at the head of the "Specs" edge, indicates that the tasks "Choose Site" and "Choose Vendor" cannot be started until after "Specs" has been completed. They can, however, be done at the same time. A node with two incoming edges, such as the one at the tail of the "Install" edge, indicates that the tasks indicated by the incoming edges are all prerequisites to the task(s) of any outgoing edges.

Assume that this satellite dish is to be an uplink as well as a downlink dish, so that a television studio is also necessary. This would involve adding five additional tasks.

T_7: Choose a studio equipment vendor
T_8: Purchase studio equipment
T_9: Prepare the studio space
T_{10}: Install studio equipment
T_{11}: Test studio and dish connections

These would also involve additional prerequisite relationships.

T_1 must be completed before T_7 can be started.
T_7 must be completed before T_8 can be started.
T_1 must be completed before T_9 can be started.
T_8 and T_9 must be completed before T_{10} can be started.
T_6 and T_{10} must be completed before T_{11} can be started.

Redraw the network diagram adding these additional tasks and prerequisites.

TASK 2: CRITICAL PATHS

Suppose each of the six jobs in the original network requires a number of weeks for completion as indicated in the following chart.

Task	**Required Time (weeks)**
T_1	3
T_2	3
T_3	2
T_4	2
T_5	4
T_6	1

The required times can be added along each path from S to F in the network. For instance, along the path from T_1 to T_2 to T_4 to T_6, the total time necessary is 9 weeks. This means that the project can not be completed in less than 9 weeks. In this network, there is also a second path from S to F, the path from T_1 to T_3 to T_5 to T_6. This path has total time of 10 weeks. This indicates that the total project needs at least 10 weeks. The project then requires as much time as the path of greatest total time (10 weeks in this case). This path is called the *critical path*.

One use of the critical path arises when the contractor wishes to shorten the project. If a task along the critical path (say T_5) is shortened by 1 week, the total time of the project also is decreased by one week. This process is called *crashing* and will be returned to shortly. Notice however, that shortening T_5 by 2 weeks only shortens the project by one week, since the other path (of length 9) becomes the critical path once T_5 is crashed.

Returning to the problem of the bathroom renovation, assume that the project involves the following activities, times and costs.

ACTIVITY TASKS	**ACTIVITY DURATION**	**ACTIVITY COSTS**
Demo old bathroom	2 days	$1,000
Purchase plumbing fixtures	10 days	$1,500
Rough plumbing	3 days	$1,200
Purchase light fixtures	5 days	$700
Rough electrical	2 days	$1,200
Finish plumbing	1 day	$600
Finish electrical	1 day	$600
Rough carpentry	3 days	$1,500
Finish carpentry	3 days	$1,500
Painting	2 days	$500
Tile	2 days	$800
Purchase accessories	1 day	$300

Assume also that the project can be represented by the following network diagram.

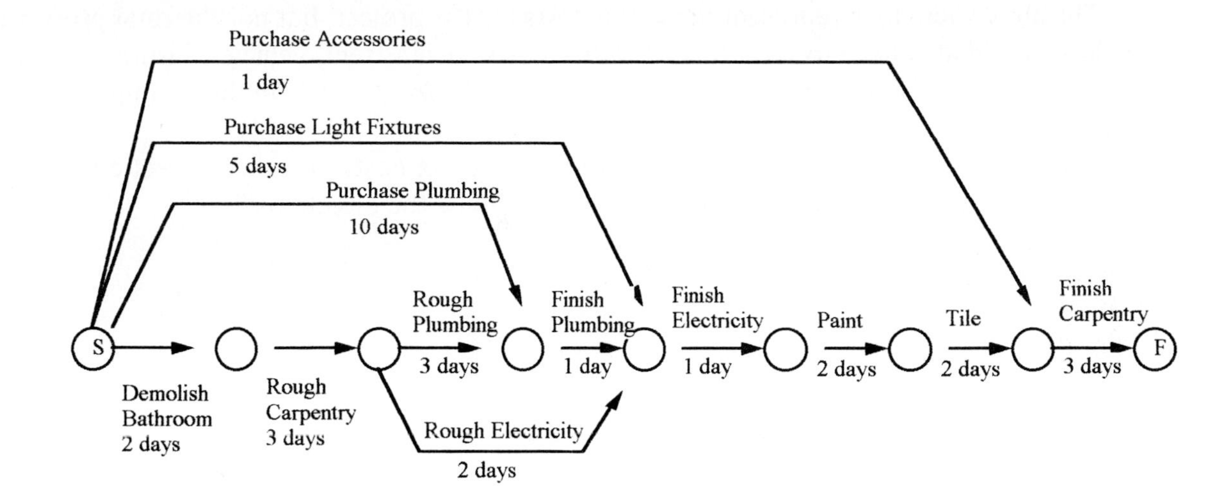

Calculate the lengths in days of all the paths from S to F for this project. (You may wish to label the nodes or the tasks so that you can more easily represent the paths).

Path	Length of Path (in days)

Which of these is the critical path?

TASK 3: OPTIMIZING PROJECT COSTS

The above activities represent the *direct costs* for the project, but not the total project cost. The builder would also have to cover his/her office costs such as rent, vehicles, insurance and other miscellaneous costs. These costs are called the project's *indirect costs*. In this example, assume an indirect cost of $250 / day.

The *total cost* of the project is the sum of all the direct costs plus the indirect costs. Using your critical path as the length of the project, what would be the direct cost?

What would be the indirect cost?

What would be the total cost?

Before proceeding with the project as estimated and scheduled above, project managers should next verify that the project is organized in the most optimum fashion. Note that the indirect cost is expended in a linear fashion; that is, for every day the project runs the company expends $250. Likewise, if the project can be shortened a day the company will save $250.

As mentioned earlier, the process of shortening a project is called *crashing*. As long as the cost to shorten a project by a specified time interval is less than the project's indirect cost during that interval, then the project is moving towards a more optimum duration. To begin the optimization process a project manager must determine how much it will cost to shorten each activity, along with the minimum possible activity duration. As activity durations are shortened the direct cost of the activity increases. This is because the activity is being accelerated and is no longer being done as efficiently as first planned. Accelerated work may require overtime pay, a crowded work place, additional equipment and other drastic arrangements, but again this acceleration makes sense if it shortens the project and the cost to accelerate is less than the indirect cost of the project. Assume the following crash costs for the bathroom renovation:

Activity	Minimum Duration Possible (days)	Additional Cost Per Day Shortened
Purchase Plumbing Fixtures	7	$150
Finish Carpentry	2	$175
Rough Carpentry	2	$150
Paint	1	$400

Compute the values for indirect cost, direct cost and total cost for each length of time listed under project duration. Record the results of these calculations in the Optimization Table below. Be sure to allow enough time for all activities to be completed.

When selecting activities to crash, work 1 day at a time, selecting the least expensive first. Also be sure that the activity chosen lies on a critical path. (You may wish to refer to the earlier discussion of crashing in the satellite dish example.)

OPTIMIZATION TABLE

Activity to be Crashed	Revised Duration	Indirect Cost	Direct Cost	Total Cost
None	19			
	18			
	17			
	16			
	15			
	14			
	13			

Can this project be done in any shorter duration? Why or why not?

From the optimization table, plot the direct cost (on the vertical axis) versus the project duration (on the horizontal axis) on the first set of axes. On the second set of axes, plot the indirect cost versus the duration and, on the third set, the total cost versus the duration.

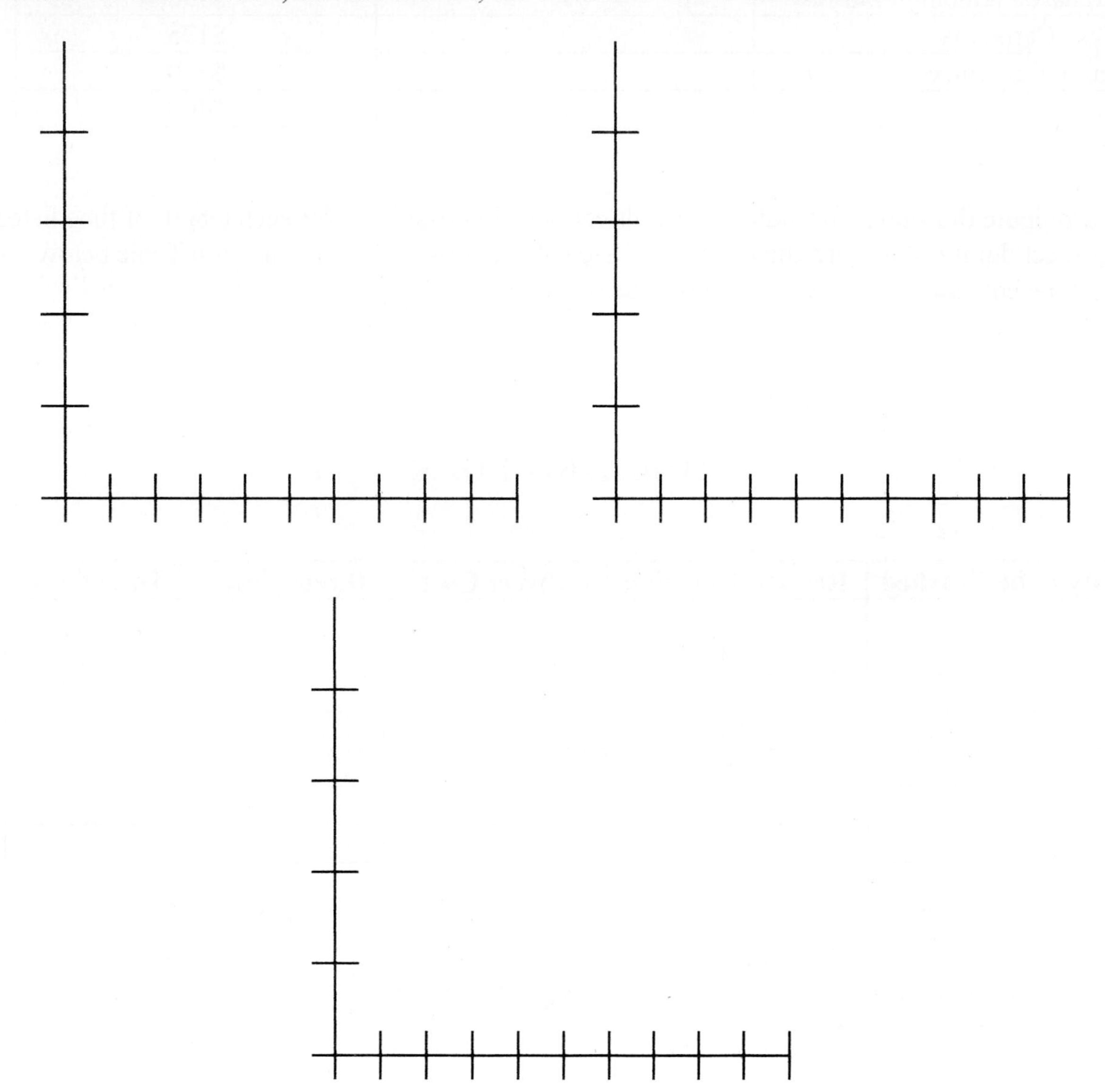

The optimal cost for the project would be the minimum total cost that can be achieved by crashing zero or more activities. What is the optimal cost for this project?

Suppose the employer offered a $200/day bonus for every day that the job was finished before the promised deadline. If you originally promised that the job would be finished based on the original (uncrashed) critical path, every day saved now nets you a $450 savings ($250 indirect costs plus $200 bonus). What would be the optimal project duration and project cost under these conditions?

III. OPTIMIZATION OF CONTINUOUS FUNCTIONS

Another type of optimization occurs when selecting a cutting speed for a high speed cutting tool. The speed must be chosen so that there will be a long tool life and a high metal removal rate. Since the cutting speed is any number (rather than an integer). This is an example of the optimization of a continuous function, as opposed to a discrete function.

As in the previous section, there are different criteria which could be used in the optimization. It might be appropriate to find the cutting speed which minimizes the time spent cutting each piece. It might also be appropriate to minimize the cost of cutting each piece. Both criteria will be considered here.

TASK 4: MAXIMIZING THE PRODUCTION RATE

Three elements contribute to the total production cycle time for one cutting tool:

a) Part handling time, T_h : This is the time it takes the operator to load the part into the machine tool at the beginning of a production cycle and to unload it at the end of production.

b) Machining time, T_m : This is the time that the tool is actually engaged during the cycle.

c) Apportioned tool change time, T_c : At the end of the tool life, the tool must be changed. The time it takes to change the tool must be distributed over the number of parts cut during the tool life.

The total time, T, which can be attributed to cutting one part is then the sum of these three quantities.

For a specific example, consider the problem of using a high speed steel cutting machine to turn a 1.50 inch long unhardened steel rod from a diameter of 1.00 inches down to a diameter of 0.90 inches. The part handling time, T_h, will be assumed to be 5.0 minutes. The machining time, T_m, in a straight turning operation is given by

$$T_m = \frac{\pi DL}{vf}$$

where D is the diameter of the rod (1.00 inches), L is the length of the rod (1.50 inches), f is the feed speed, and v is the cutting speed. Keeping v as a variable, assume that f is 0.010 inches/revolution.

Simplify T_m as an equation in v.

The apportioned tool change time can be found by dividing the actual time needed to change the tool, T_t ,by the number of parts which can be cut during the tool's lifespan. If this number is called n_p, the formula for the apportioned tool changing time becomes

$$T_c = \frac{T_t}{n_p}$$

The amount of time necessary to change the tool is a constant, 2.0 minutes in this example. The number of pieces per tool, n_p, is simply the lifespan of the tool, T_L , divided by the time neccessary to turn one rod, T_m.

$$n_p = \frac{T_L}{T_m}$$

The lifespan of the tool, T_L, is a function of the cutting speed, the depth of the cut, and the materials involved. For this example, it can be shown experimentally (by the *Taylor Tool Life Equation*) that

$$T_L = \left(\frac{200}{v}\right)^8$$

Using this formula as well as your formulas for T_m and n_p , **write a formula for the apportioned tool change time, T_c, as a function of v.**

The formula for the time attributable to turning a single rod then becomes

$$T = T_h + T_m + T_c$$

Write this function as a function of v and graph it on a calculator or other graphing utility. For what value of v is this time minimized?

If the operator could be trained to load and unload the part in a shorter time, how would you expect these values to be affected?

TASK 5: MINIMIZING THE COST PER PART

The cost of turning the rod in the previous example comes from two sources, material and labor. The cost due to labor is the cost of paying someone to run the machine. Assuming a fixed cost of \$30.00 per hour (or \$0.50 per minute, since the units are all in minutes up to this point), the labor cost of turning one rod becomes:

$$C_L = 0.50T_m$$

The materials cost for turning one rod becomes the cost of the tool (including replacement, regrinding etc.) divided by the number of rods milled by the tool before replacement. Assuming a tool cost of \$3.00, the materials cost then becomes

$$C_M = \frac{3.00}{n_p}$$

yielding a total cost of

$$C = 0.50T_m + \frac{3.00}{n_p}$$

Using your results from Task 4, write this function as a function of the turning speed, v.

Graph this function on a graphing calculator or other graphing utility. For what value of v is this cost minimized?

MATHEMATICS LABORATORY INVESTIGATION

DESIGN OF SPIRAL AND CIRCULAR STAIRS

Topics: **RIGHT TRIANGLE TRIGONOMETRY**

Prerequisite: *Definitions of Trigonometric Functions*

I. INTRODUCTION

In design, curved forms suggest openness and free space. In the built environment, round or curved spaces have a special character that tends to draw your eye to them. Spiral and circular stairs incorporate these curved forms to suggest movement as well as openness and variety. Through the use of spiral and circular stairs, the designer can seize the opportunity to make a grand statement while still preserving the utility of linking different levels of a space.

While spiral and circular stairs can be very efficient in that they can take up a small amount of floor space (the *footprint*), they do have the potential for causing accidents. Designers therefore need to pay close attention to codes and guidelines for safety and comfort.

II. DESIGN

The stair treads of a spiral staircase can be visualized either as a sector of a circle or as an isosceles triangle, depending on whether the outer edge is curved or not. (see figure 1)

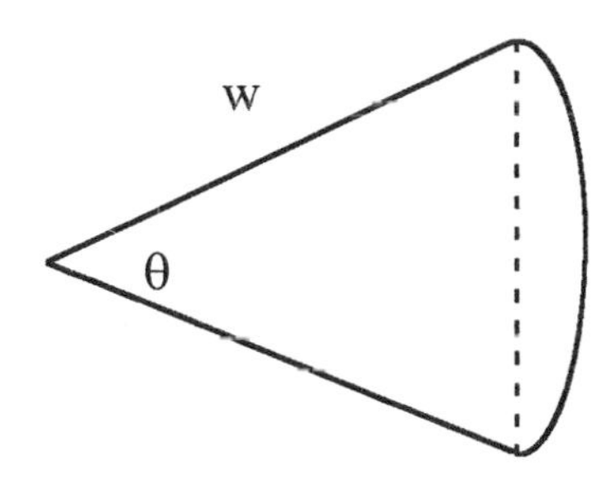

figure 1

In this investigation, it will be assumed that the outer edge of the stair is straight, but in any case, the quantity, w, (as shown above) is referred to as the *tread width* and θ is referred to as the *central angle*. Staircases are often referred to indirectly by means of their central angle. For example, if the central angle is 30°, it would take 12 steps to complete one 360° traversal of the staircase. Such a staircase is often referred to as *12 stairs around.*

III. STAIRCASE CONSTRUCTION

TASK 1: Spiral Staircase Construction

For both comfort and safety, building codes mandate the minimum depth of a stair tread. For spiral staircases, tread depth is not a constant, but increases the farther away one moves from the central axis. One way of defining the minimum depth is to say that it must be at least 8 inches deep from the riser to the front of the tread when measured at a distance of 15 inches from the inner edge of the stair. (see figure 2)

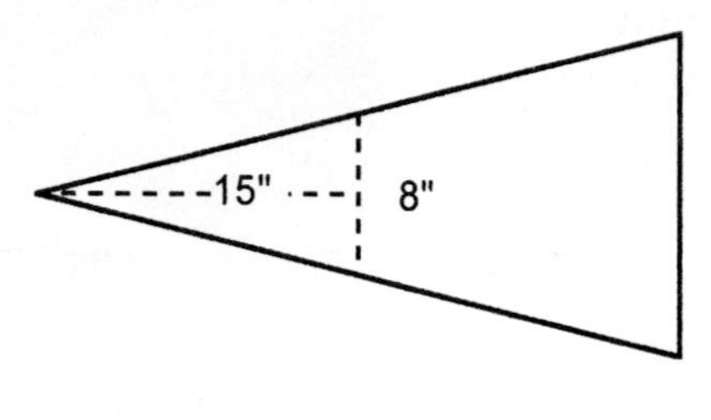

figure 2

What is the minimum central angle allowable for a staircase which complies with this code?

Based upon this minimum central angle, what is the minimum number of "stairs around" (as defined in II above)?

Building codes also specify the minimum head clearance for spiral staircases. One such Massachusetts code (MSB 816.2.2) states that the headroom in all parts of a stair enclosure should be at least 6'8" . A guideline from *Time Saver Standards for Interior Design and Space Planning*, by DeChiara, Panero and Zelnik states that the headroom on circular stairs should be calculated on the basis of three quarters of a circle. (For example, if the staircase is 20 stairs around, the height gained by climbing 15 stairs should be at least 6'8".) Using your results for the minimum number of stairs around and the minimum central angle, what should the minimum height of a step be in order to comply with this code?

Photo courtesy of Boston Athenaeum

For stairs in general, a step height of seven inches is considered the most comfortable. How many stairs around would be needed in order to achieve the head clearance specified above?

Write a paragraph summarizing your results from Task 1.

TASK 2: Circular Stairs

A common staircase is 16 stairs around. This is because (as you have already determined), it gives a comfortable step height of under 7". What is the central angle for a stair tread on such a staircase?

The central angle you determined should have come out too small to satisfy the code described at the beginning of Task 1 in that at a distance of 15" from the central axis, the depth of a stair tread will be less than the required 8". (See figure 2 above) At what distance from the central axis will the depth of the stair tread reach the required 8"?

Why do you think this 8"/15" code exists?

In order to satisfy the code, you could cut out enough of each stair tread so that the depth reaches 8" within the required 15" limit. (see figure 3)

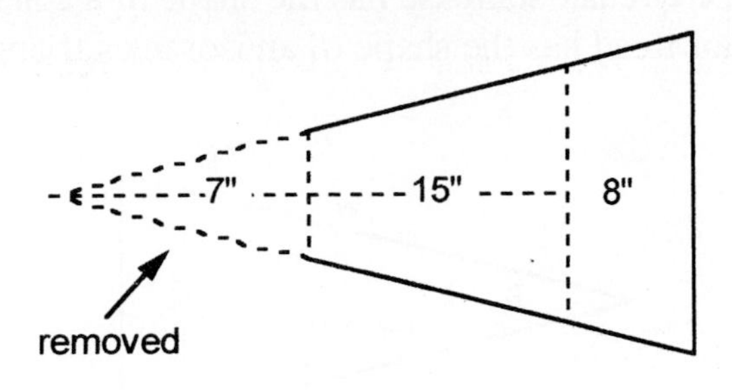

figure 3

For example, if the stair tread did not reach 8" until a distance of 22" from the central axis, you would cut the first 7" off the narrow end of each stair tread. The shape of each tread would then be an isosceles triangle with a small isosceles triangle removed from the end near the central axis. The effect of this is to create a well in the middle of the staircase. This type of staircase is called a *circular* (as opposed to spiral) *staircase.*

If you had an 8' square space in which to build a spiral or circular staircase which is 16 stairs around, how much would have to be removed from the end of the stair near the central axis to bring it within the 8"/15" code? (This is equivalent to calculating the radius of the center staircase opening.)

TASK 3: Handrail Length

The handrail for a spiral or circular staircase has the shape of a spiral. In order to calculate its length, recall that a stair tread has the shape of an isosceles triangle

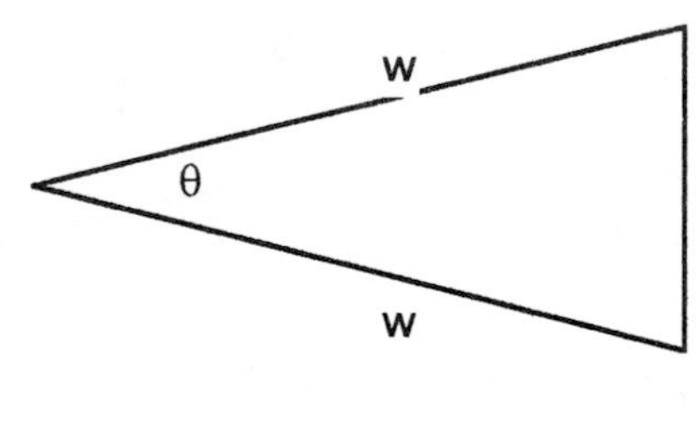

figure 4

where w is the stair width (measured from the center) and θ is the central angle. (In the case of a circular staircase, the inner part of this stair will be removed, but for this Task, w will still be measured from the central axis.)

Look at the following picture of several stairs.

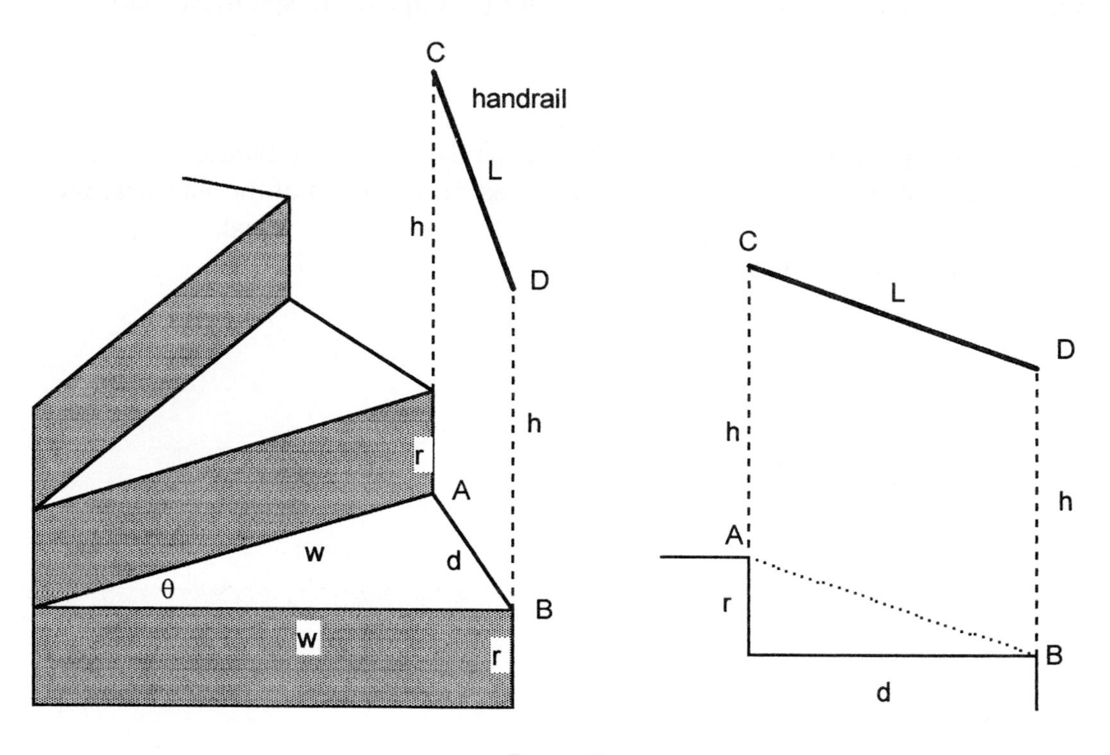

figure 5

Here h is the height of the handrail, r is the height of a riser, d is the length of the outer edge of the stair tread and L is the length of one section of handrail.

As a first approximation, assume that each section of the handrail is actually straight. Write a formula expressing d in terms of w and θ.

Write a formula for L in terms of the other variables.

If you have circular stairs which are 16 stairs around and have 7" risers and the radius, w, is 3 feet, approximate the length of handrail needed to make one full revolution about the central axis.

The actual shape of the handrail is a helix (spiral).

Photo courtesy of Boston Athenaeum

The formula for the exact length of one turn of a spiral is given by

$$2\pi\sqrt{R^2 + K^2}$$

where R is the radius of the spiral and K is a measure of the slope of the spiral. In particular, K is the change in vertical height of the spiral through one full revolution divided by 2π.

Determine a formula for K and use these formulas to determine the length of the handrail.

This result is exact (except for rounding). How good was your first approximation? Express not only the error, but the percent error.

MATHEMATICS LABORATORY INVESTIGATION

DESIGN OF A STRAIGHT STAIRCASE

Topics: **LINEAR INEQUALITIES**
Prerequisite: *Algebra, Graphing Linear Functions*

I. INTRODUCTION:

Since ancient times, architects have been trying to design stairs with dimensions suitable to the human gait. Various rules of thumb have been employed but it was not until about 1762 that François Blondel, director of the Royal Academy of Architecture in Paris, first mathematically defined the rule of thumb that is still in use today. He observed that the normal human walking gait was about 24 inches and that this amount needed to be reduced by about two inches for every vertical inch of the stair.

Time has borne out his observation. When the Lincoln Center for the Performing Arts in New York City was built, the plaza staircase leading to the building had shallow risers (3 3/8") and deep treads (25"). This imposed such an awkward gait that numerous falling accidents occured, especially in descending the staircase. As a result, much of the staircase was replaced by a ramp system and the remaining stairs were closed off by railings.

When designing or making changes to the built environment, building codes are an essential part of the design process. The existence of building codes focuses on life safety: to ensure the proper exits, corridors, doorways to allow escape in the event of a fire or other hazard. Another major purpose is to provide the proper use of materials and finishes reducing the possibility of hazardous fumes and smoke.

There are various codes used throughout the country; these will vary from state to state. In Massachusetts, the Commonwealth of Massachusetts State Building Code, 5th Edition, 780 CMR (MSB) is the one currently in use, adopted by the State Board of Building Regulations and Standards. In addition to state and local building codes, there are regulations and laws that affect the design of spaces. These may include The Life Safety Code (published by the National Fire Protection Association), the Architectural Barriers Board and the Americans with Disabilities Act (ADA), which is not a code at all, but a comprehensive civil rights law governed by the Justice Department. The section of the ADA that pertains to designing the built environment is the Federal Register, Part III, 28 CFR.

Remember that codes and laws are written as minimum standards. It is up to the designer to create the best solutions from these requirements and not to rely solely on providing the minimum standard.

II. BUILDING CODES

Modern building codes are designed to provide a mathematical basis for appropriate stair construction. The following are example requirements for interior, residential stairs:

Code A: (MSB 816.4.1) For residential interior stairs, the height of the risers cannot exceed eight and one quarter inches.

Code B: (MSB 816.4.2) All risers must be approximately the same height. Adjacent risers can differ by no more than 3/16" and the difference between the shortest and tallest riser can be no more than 3/8".

Code C: (MSB 816.4.1) For residential interior stairs, the depth of a stair tread must be at least nine inches.

Code D: (MSB 816.4.2) All treads must be approximately the same depth. Adjacent treads can differ by no more than 3/16" and the largest and smallest tread can differ by no more than 3/8".

Refer to the following picture for the definitions.

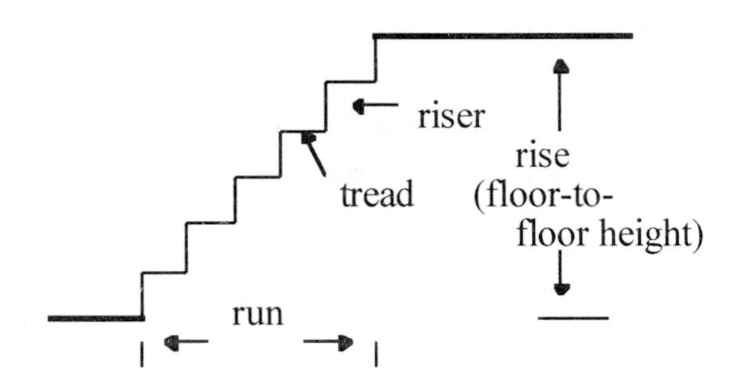

figure 1

Additionally, there are standards for comfort (which are **not** legal requirements or codes). One such standard, which, in this investigation, will be referred to as Blondel's Rule is:

Blondel's Rule: (*Architectural Graphics Standards*, by Ramsey and Sleeper) The height of a riser plus twice the depth of a stair tread should be between 24 and 25 inches.

III. STAIR CONSTRUCTION

TASK 1: Graphically Interpreting the Codes

Let R represent the riser height and T represent the tread depth. Rewrite codes A and C as inequalities.

Plotting R on the vertical axis and T on the horizontal axis, shade in the region of the first quadrant representing the dimensions allowable under these codes (the *feasible region* for riser height and tread depth).

Label any corners of this region and explain their significance.

TASK 2: Constructing a Staircase

You are to construct a residential, interior straight staircase with a floor-to-floor height of 11 feet. Using the building codes (A-D), calculate the minimum number of risers necessary and the height of each riser. Indicate also the number of treads.

Minimum Number of Risers: ________

Riser Height: ____________________

Number of Treads: ________________

For the staircase you just described, using Blondel's Rule, determine the range of tread depths. What does this say about the run of the staircase? Is the range of tread depths consistant with the building codes (ie. is it in the feasible region from Task 1)?

Tread Depth Range: ___________________

Run Range: _________________________

Use one more riser than you used in your previous example. Recompute the riser height for this staircase. Use Blondel's Rule once again to determine the range of tread depths and the corresponding range for the staircase run.

Number of Risers: __________________

Number of Treads: __________________

Riser Height: ______________________

Tread Depth Range: _________________

Run Range: _______________________

Is this answer consistant with the building codes?

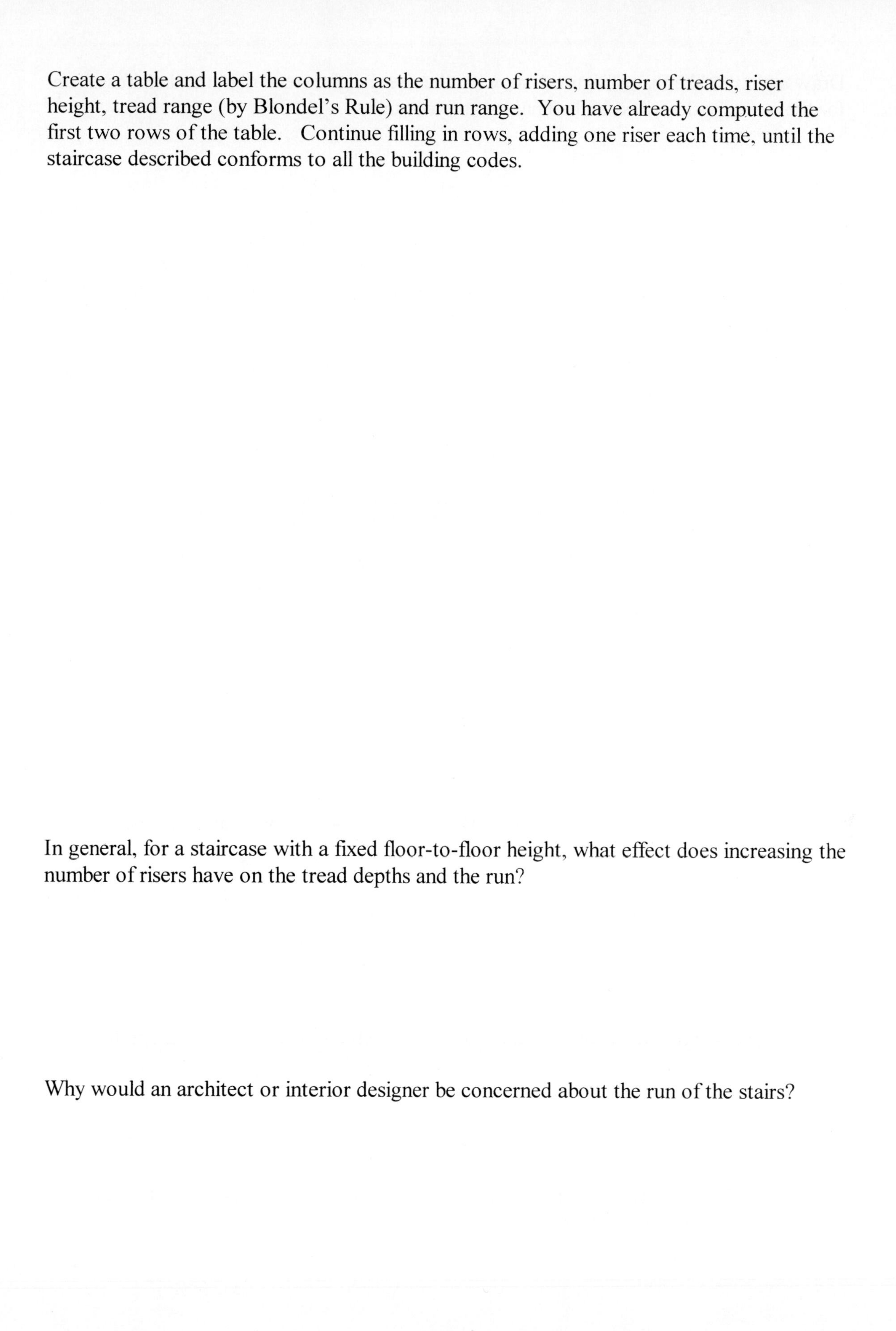

Create a table and label the columns as the number of risers, number of treads, riser height, tread range (by Blondel's Rule) and run range. You have already computed the first two rows of the table. Continue filling in rows, adding one riser each time, until the staircase described conforms to all the building codes.

In general, for a staircase with a fixed floor-to-floor height, what effect does increasing the number of risers have on the tread depths and the run?

Why would an architect or interior designer be concerned about the run of the stairs?

Draw a picture like that of figure 1 showing the riser height, tread depth and staircase run for the staircase described in the final row of your table.

TASK 3: Graphical Solutions

Using the variables from Task 1, express Blondel's Rule as algebraic inequalities.

Plotting riser height on the vertical axis and tread depth on the horizontal, graph the inequalities for Codes A, C and Blondel's Rule and shade the region which corresponds to the allowable dimensions.

Plot your solution to Task 2 in the shaded region. Label the corner points of this region and write a sentence explaining the significance of each.

TASK 4: Designing a Staircase

A new home is being designed in which there is to be a straight staircase with floor-to-floor height of 10'3" and a run of 13'. How would you recommend the stairs be constructed? (ie. How many stairs? Riser height? Tread depth? etc.)

Extensions:
1. Measure the stairs at your school or measure stairs that are uncomfortable to walk upon. What about these measurements makes the stairs comfortable or uncomfortable to walk upon? Do they satisfy Blondel's Rule?
2. The stairs you designed in Task 4 are to be carpeted. How many linear feet of carpeting would be necessary for this staircase? In general, how much carpeting is necessary for a straight staircase (in linear feet of carpet)?

MATHEMATICS LABORATORY INVESTIGATION

ELECTRONIC CIRCUIT DESIGN

Topic: **COMPOSITE FUNCTIONS**
Prerequisite knowledge: *Linear Equations, Functional Notation*
Equipment required: 9-volt battery or power supply, small breadboard, digital multimeter or voltmeter, 1N914 diode with protective sheath or straw, 10KΩ resistor, red and black alligator clips, two wire leads, ice water, thermometer

I. INTRODUCTION

A *transducer* is a general classification for any device that converts one form of energy into another form. One type of transducer is a *silicon diode*, which responds to changes in temperature. It will produce an electrical voltage readout whose value depends on the temperature of the environment of the diode.

The voltage output of the diode decreases linearly as the temperature increases. However, the practical use of such a diode in a computer-based data acquisition system requires that this output be converted to a linearly increasing voltage, preferably with an output in direct proportion to the temperature being measured.

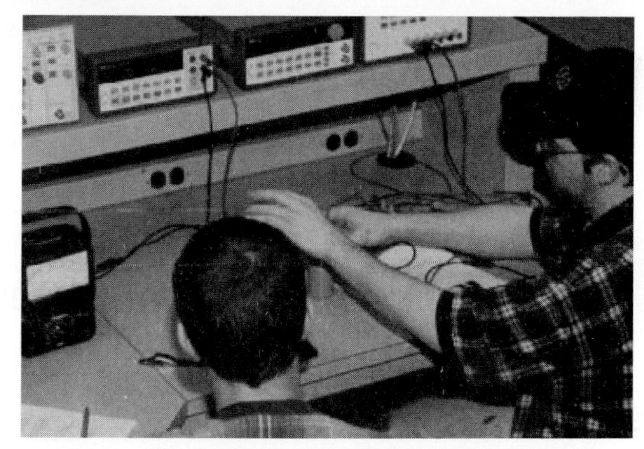

In this exercise, you will determine the mathematical properties of a typical silicon diode used as a temperature sensor, as well as the mathematical properties that would be necessary in the design of the *Signal-Conditioning Circuit* (SCC) that is required in order to convert the *diode voltage* V_D to a simple *output voltage* V_O directly proportional to the temperature T. **The objective is to find an intermediate function that will convert the decreasing $V_D(T)$ equation into an increasing $V_O(T)$ equation:**

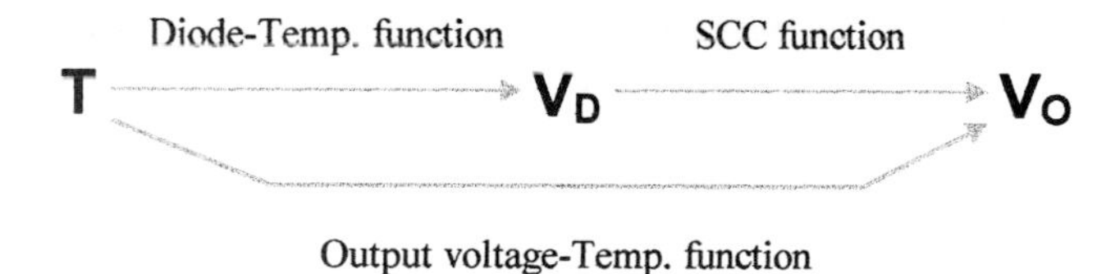

II. MATHEMATICAL CHARACTERISTICS OF THE DIODE

Set up a simple series circuit containing the diode, resistor, and 9-volt battery or power supply, as shown in the Appendix. Familiarize yourself with the operation of the meter by attaching it across the battery and verifying that it reads 9 volts. **With what precision does your meter allow you to measure voltage?**

____

Using an ordinary thermometer, measure the air temperature T_1 (in degrees Celsius, °C) in the region of your workspace, and record it below. Then attach the meter across the diode, and record the diode voltage V_{D1} (in volts) at room temperature. This will provide one set of values (T_1,V_{D1}) to help you determine an equation relating temperature and voltage for the diode.

Now, hold the diode between your thumb and forefinger. (If the diode is covered by a plastic sheath, ask for help in locating it.) Record the new diode voltage V_{D2}, as well as your skin temperature T_2, so you will have a second set of values (T_2,V_{D2}).

Finally immerse the diode, in its protective sheath, in ice water, again recording temperature and voltage at this third data point (T_3,V_{D3}).

Temperature T (°C)	**Diode Voltage V_D** (volts)

TASK 1: On the next page, **draw a well-labeled graph** which represents the diode voltage V_D as a function of temperature T. Assume that you are interested in the temperature range 0°C - 100°C only (which covers the temperature range of liquid water), and that voltage and temperature are linearly related over this range.

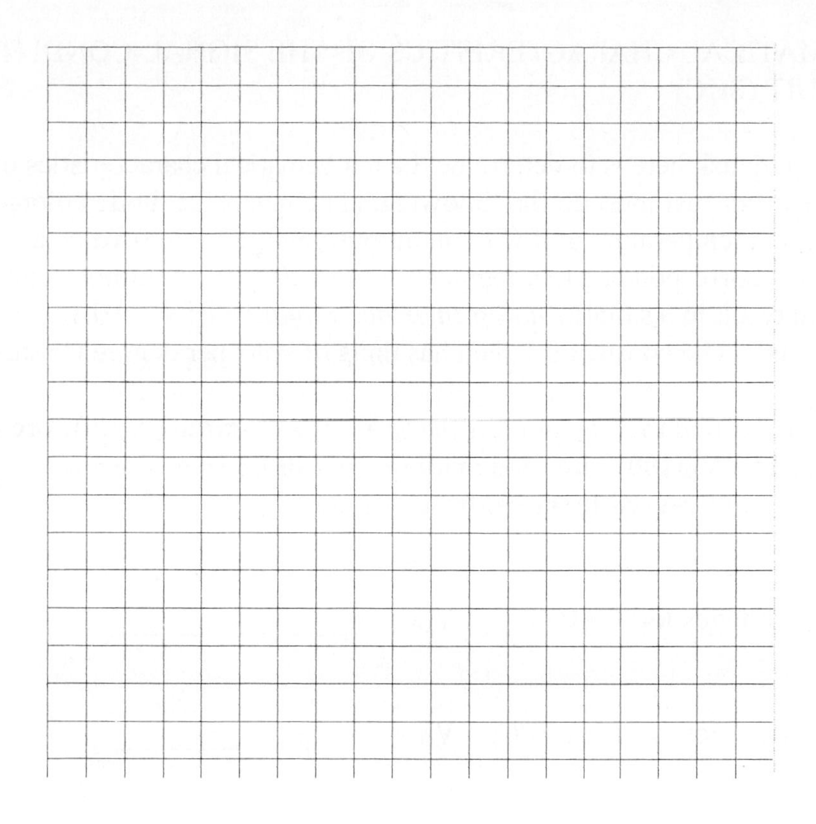

Determine the function $V_D(T)$ relating diode voltage and temperature by finding the equation of the *line of best fit* for your data. **Write the equation here,** and also **draw in the line on the graph.**

____

What are the *units* of the slope of this graph, and what is the *physical meaning* of the slope?

____

____

____

You should use your calculator to plot the three data points and the graph of your $V_D(T)$ function, and verify that your equation is correct.

III. MATHEMATICAL CHARACTERISTICS OF THE SIGNAL-CONDITIONING CIRCUIT (SCC)

The design task here is to determine the mathematical characteristics of a *Signal-Conditioning Circuit* that must do the following: (1) convert the diode voltage V_D corresponding to a temperature of 0°C to an output voltage V_O of 0 volts, and (2) convert the diode voltage corresponding to a temperature of 100°C to an output voltage of 10 volts. This will result in a simple *Analog Interface Equation* of $V_O = 0.1T$, with V_O measured in volts. (The constant 0.1 then has units of volts per degree Celsius, or V/°C.)

TASK 3: To find the equation for the SCC, **first compute $V_D(0)$, the diode voltage at 0°C, and $V_D(100)$, the diode voltage at 100°C, from the equation you found in part II, and record them below as V_{D1} and V_{D2}.**

Voltages for T = 0°C: V_D = _______________ V_O = 0 volts

Voltages for T = 100°C: V_D = _______________ V_O = 10 volts

TASK 4: **Plot the pairs of voltage values** as points on a well-labeled graph of output voltage vs. diode voltage, and draw in the line representing V_O as a function of V_D.

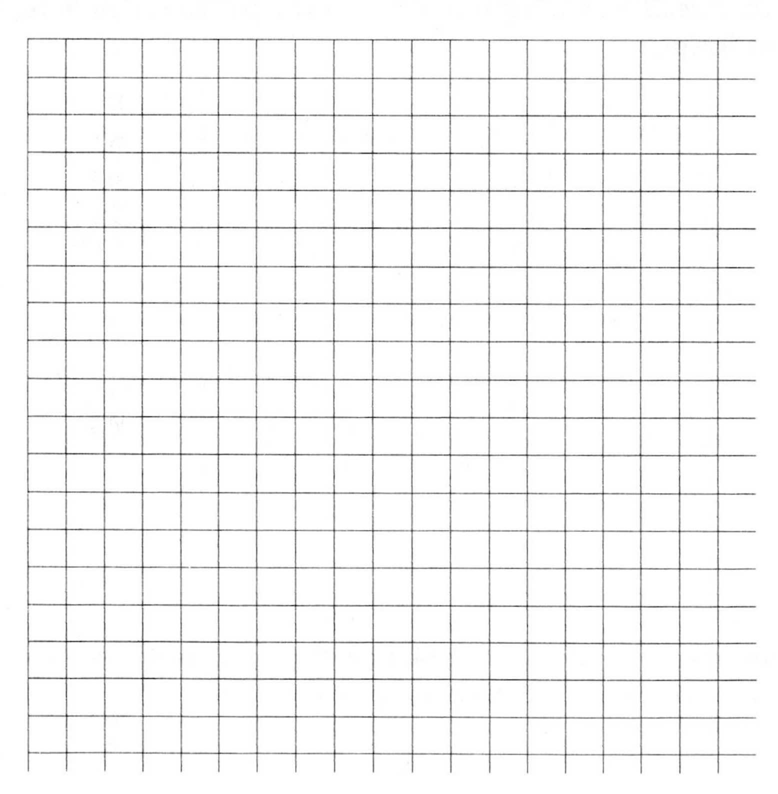

TASK 5: Finally, **derive algebraically an equation giving output voltage as a linear function of diode voltage $V_O(V_D)$.**

Enter this second function into your calculator (keeping $V_D(T)$ from part II) and again check by graphing this function to verify that it contains the points that you plotted. If it doesn't, discuss with your group and recalculate.

IV. A COMPOSITE FUNCTION

You now have three related variables: temperature T, diode voltage V_D, and output voltage V_O. A good way of comparing them is to list values of all three variables side by side.

TASK 6: Using your calculator, create the variable lists as follows:

(1) Key in values of T in the first list, called L_1, as {0,10,20,...100};

(2) Next, enter L_2 as your function $V_D(T)$ to display diode voltage values that correspond to the temperature values in L_1;

(3) Now, write the expected values of V in the third column L_3 (that is, the values which will provide the desired output range of exactly 0-10 volts).

(4) Finally, enter L_4 as your function $V_o(V_D)$ to display the corresponding output voltage values. Write down the three lists below.

Complete the table on the next page, or include a printout of your calculator lists.

T	V_D	expected V_O	actual V_O
0			
10			
20			
30			
40			
50			
60			
70			
80			
90			
100			

Discuss with your group any discrepancies you find between the two columns of V_O values, and explain them here:

By comparing the three lists you have created, you can see how temperature values are changed first, by the diode, into voltage values, and then how the signal-conditioning circuit would change these into simple output voltages for use by a computerized data recording device. You can also see by inspection that the output voltage list gives values of the function $V_O(T) = 0.1T$, with the first (temperature) list as inputs. (It is the job of an electronics engineer to actually design the circuit that would accomplish these transformations.)

TASK 7: Mathematically, we say that V_o is a *composite function* of T, because it is actually made up of <u>two</u> functions: the <u>output</u> of the function $V_D(T)$ becomes the <u>input</u> of $V_O(V_D)$. You may be able show this in another way by forming the composite function directly on your calculator. Try replacing the input variable of your function representing $V_O(V_D)$ with the entire function representing $V_D(T)$. (On some calculators, this is accomplished by replacing the X in $Y_2(X)$ with $Y_1(X)$. **Graph this function, and describe here what you see:**

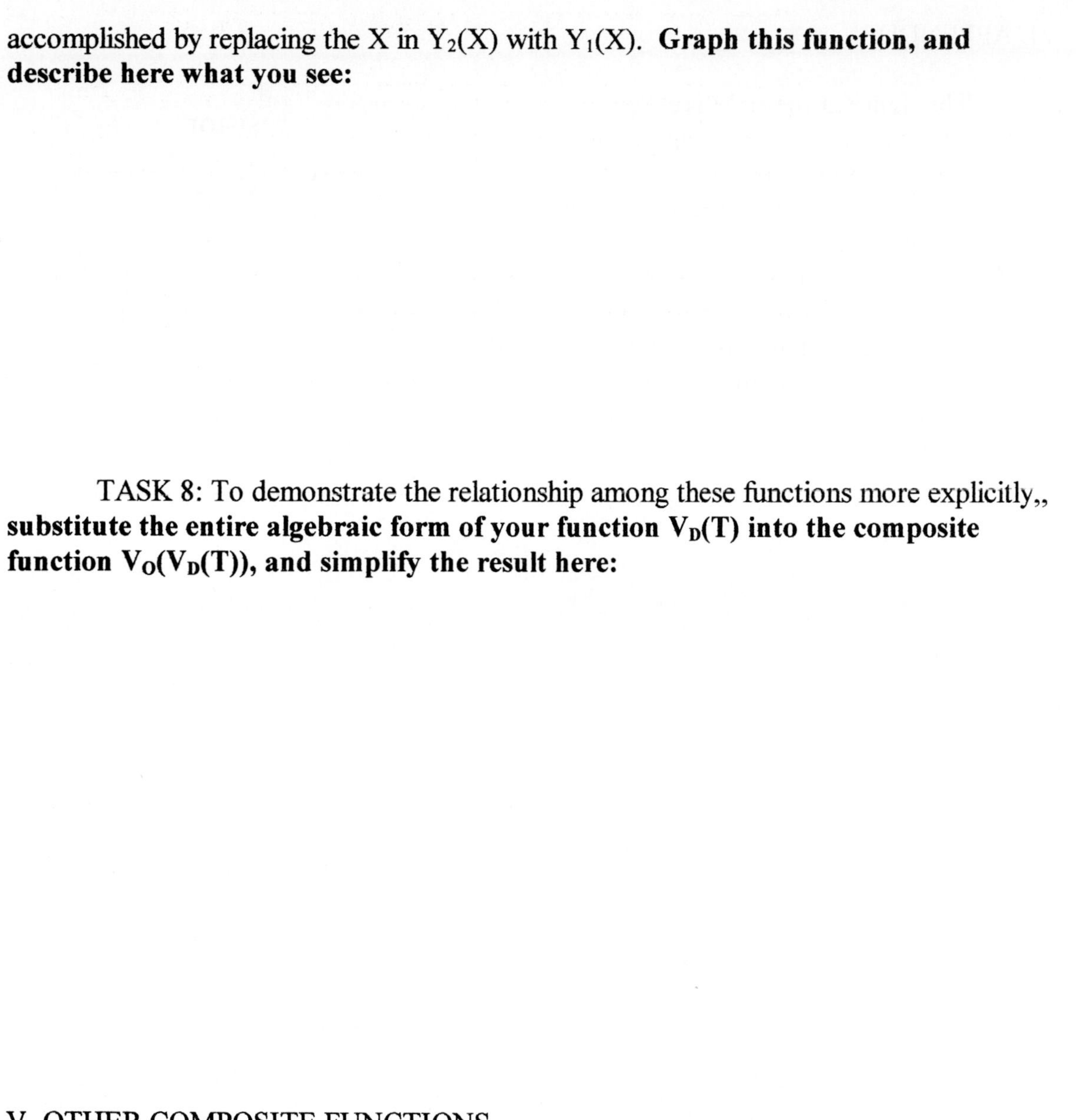

TASK 8: To demonstrate the relationship among these functions more explicitly,, **substitute the entire algebraic form of your function $V_D(T)$ into the composite function $V_O(V_D(T))$, and simplify the result here:**

V. OTHER COMPOSITE FUNCTIONS

Describe below any other examples you can think of that involve composite functions: that is, situations in which an "output" variable depends on the variable of another variable, which in turn is determined by the value of a third variable.

VI. APPENDIX

The figure at the right is called a *schematic diagram* of the circuit used to collect data in this laboratory investigation. It represents a *series circuit* containing a battery, resistor, and diode. A series circuit is one in which all components are connected in a single unbranched chain, so that the same current flows through every part of the circuit.

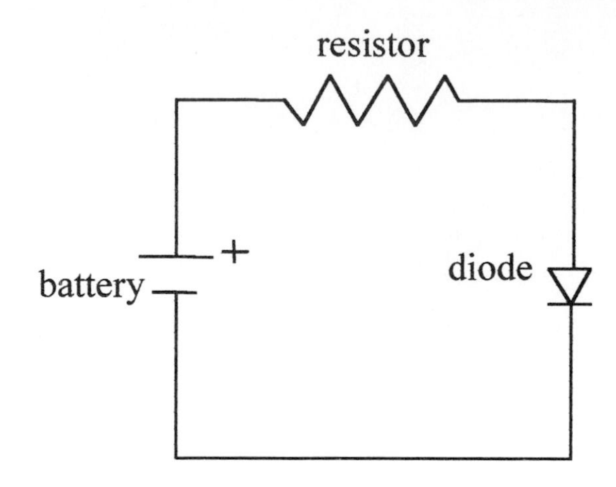

One of the simplest ways of accomplishing this circuit construction is to use a *resistor block*, which minimizes the number of wires needed. A diagram of a possible setup is shown below. A key requirement is that the negative (black) terminal be connected to the negative (black) lead of the diode, since a diode is *directional* and only allows current to pass in one direction.

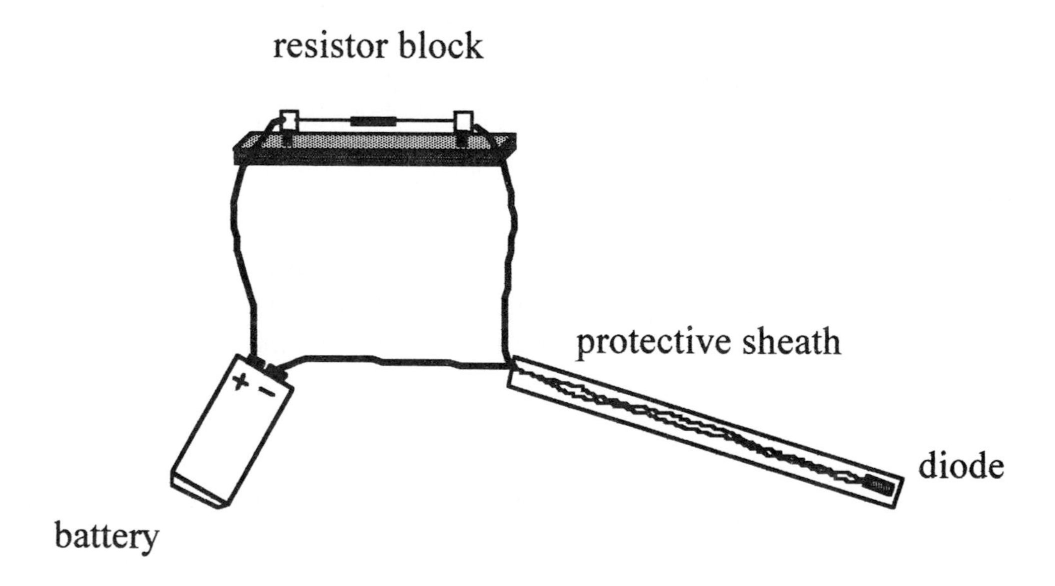

MATHEMATICS LABORATORY INVESTIGATION

ENGINEERING MEASUREMENT

Topics: **SCIENTIFIC NOTATION, ACCURACY, PRECISION, AVERAGE**
Prerequisites: *Metric System, perimeter, area, volume*
Equipment: meter stick, micrometer, vernier caliper, wood block

I. INTRODUCTION

Micrometer

When mechanical parts are produced to be put into any manufactured product on an assembly line or in a repair shop we expect that they will fit. This requires the ability to make accurate measurements, and to keep dimensions on manufactured parts within close tolerances. No matter how careful one is, all measurements are approximations limited by the equipment, the variability in the production process of the item to be measured, and the skill of the operator.

Different situations require different measurement precision. In building a house a carpenter usually measures to 1/16 inch (1.6 millimeters). A cabinet maker needs to make measurements to 1/32 inch (0.8 millimeters). In a machine shop basic measurements are expected to be within 0.0001 (1/1000) inch (0.02 millimeters). When precise fits are required between mating parts, closer tolerances are required.

When parts fit poorly, performance is compromised. For example, every few months a particular automobile would not start and replacing the starter had not permanently repaired the car. Since it was determined that the starter was not defective, an engineer was assigned to find the source of the problem. In that automobile the starter motor shaft and the flywheel ring gear should have been 1/16 inch apart when the starter motor was not engaged. In an assembly line part tolerances accumulate. This assembly was four times the distance required. Accurate measurements indicated that there was 1/4 inch clearance between the starter motor gear and the flywheel ring gear. A hole was drilled 3/16 inch closer to the flywheel ring gear. The problem was solved.

What types of situations have you encountered where you need to know the tolerances allowed?

II. ACCURACY VS PRECISION USING MEASURING INSTRUMENTS

A carpenter working on a house will use a tape measure rather than a micrometer because he must measure large dimensions, and he does not need the close tolerances measured with a micrometer. A machinist on the other hand uses the micrometer because his parts are smaller, and he does need the close tolerances.

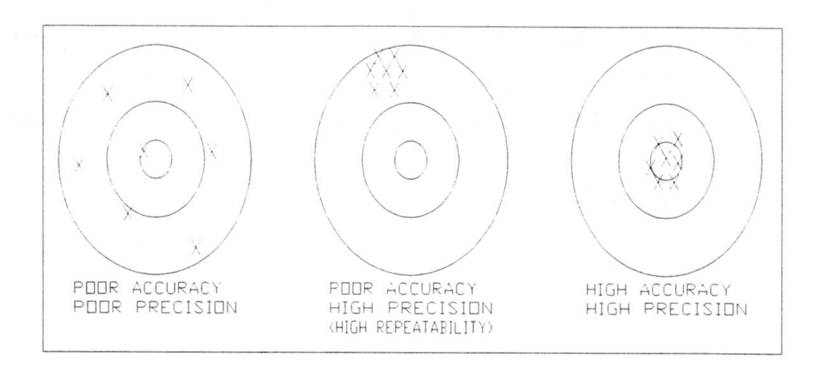

PRECISION:

A limitation to the accuracy and precision of a measurement comes from the instrument used for making the measurement. The ability of an instrument to be used in making a very refined measurement is referred to as the precision of the instrument. For example, it is more precise to use a micrometer which measures to .001 inch (or 0.01 mm) than it is to use a tape measure which measures to 1/16 inch (or 1.6 mm). The precision is also related to the repeatability of making the same measurement.[1]

ACCURACY:

If the micrometer is not initially set to zero, it may have high measuring capability (precision), but the measurements produced by using it may not be close to the actual dimension of the part. It then has poor accuracy. It must be compared to a standard to check its calibration error.

In general, measuring instruments are calibrated so that the measurement error using that device is no more than 1/2 of the smallest scale division on the instrument scale.
If, when checked against a standard, the error of an instrument throughout its range is less than 1/10 of the smallest scale division on the instrument, then the instrument can have its accuracy certified, and it can be used as an accurate measuring device. For example, if an instrument reads to .001 units, then the error is .0005 units.

[1] Jerry Lee Hall and Mahmood Naim, "Instrument Statics," in Instrumentation & Control, Chester H. Nachtgal ed, Wiley Series in Mechanical Engineering Practice, New York, John Wiley & Sons, Inc. 1990, p.62.

When length measurements are made, it is necessary to **record the result in such a way that the recorded value expresses the precision of the measurement**. If a tape measure is 12 meters long (39.4 ft) and is calibrated in millimeters, it can measure values in the following:

a. tens position, (10 meters).
b. units position (meters)
c. tenths position (decimeters)
d. hundredths position (centimeters)
e. thousandths position (millimeters)

Thus this measuring device is capable of supplying a maximum of five *significant digits*.

III. PREPARATION: Scientific Notation & Significant Digits

Assume that a length measurement has been made and the value has been recorded as 105 centimeters. This number can be written in *scientific notation.* In scientific notation the decimal point is placed after the first non-zero digit on the left and the value of the number is retained by multiplying the new decimal by 10 to the appropriate power. 105 is written as 1.05×10^2 and 0.00105 is written as 1.05×10^{-3}.

If you change the mode in your calculator from Normal to Scientific, it will give you a calculator display form of scientific notation. Most calculators do not display exponents using a superscript notation. However standard form requires that you convert the calculator display into a number using the exponent in superscript form. Under some conditions your calculator may display 0.00000105 as 1.05E-6 which is 1.05×10^{-6}. You should be able to read numbers in decimal form, by calculator display scientific, or by scientific notation and then write them using decimal form and scientific notation (using superscripts for the exponents).

Task 1: Write the following numbers in scientific notation, and as a calculator display.

Number	Calculator Display Normal	Calculator Display Scientific	Scientific Notation Standard
134006			
0.000257			
0.029035			
1000.6			
0.000507			
2^{-12}			
1370.4			

THE SIGNIFICANCE OF ZERO

Some zeros in a recorded number are significant because they indicate that a zero measurement has been made in that place value. Other zeros only hold the place value. It is sometimes confusing to distinguish the number of significant digits recorded when there are zeros in the recording.

The following are some guidelines:

Note that a zero within a number is always significant, as in 5.07, 108, and .302, (3 significant digits each);

When a number is 10 or greater, and trailing zeros are there only to locate the decimal point, then the zeroes are not significant as in 2300 and 130 (2 significant digits each);

For a decimal number smaller than 1.0, the zeroes following the decimal point before the first non-zero digit are there to locate the decimal point and are not significant, as in .00415 (3 significant digits) and .00009 (1 significant digit);

All the zeros at the end of a decimal number are significant, as in 0.91500 (5 significant digits) and 490.0 (4 significant digits). Those ending zeros indicate that the number was precisely measured to the place value of the final digit.

Most calculators do not take into account the significance of zero. You need to learn how to use the guidelines and adjust the standard form of the number in scientific notation to include significant zeros.

Task 2: Review and fill in the table below.

Notice that if you put your calculator in Scientific mode, the display will read the same for the first four examples below but the scientific notation will be different for each of them.

Number	Significant Digits	Scientific Notation	Number	Significant Digits	Scientific Notation
5.00	3	5.00×10^0	0.0501		
500.00	5	5.0000×10^2	41300		
500.	3	5.00×10^2	0.000450		
500	1	5.00×10^2	1370.0		
.000345	3	3.45×10^{-4}	64000230000		
.0350	3	3.50×10^{-2}	7.035×10^{-3}		
48.005	5	4.8005×10^1	3.6900		

If a measurement made of the distance to the moon using a laser has a maximum error of 1 mile, and the distance to the moon is about 240,000 miles, how many significant figures would be recorded in the result?_______________

IV. AVERAGING MEASUREMENTS

Because objects being measured often have some variability, it is necessary to make repeated measurements of the dimension or other variable, such as temperature or pressure, being measured. The *recorded average* of your measurements should take into account the accuracy and precision for the measurements. It is different from simply averaging, which, as you might expect, is the sum of the listed data (using all the useful digits) divided by the number of data measured. The recorded average should be rounded to the decimal place for which there is no variation and will include a ± sign followed by the *maximum likely error*.

In practice there are several methods used to determine both the recorded average and the maximum likely error. Statistical methodology developed in a later laboratory investigation will provide another way to get a measure of the error.

The following five different values were found by measuring the area of a region using a planimeter: 15.58212, 15.58144, 15.58301, 15.58293, 15.58031.

Task 3: Learning a procedure for finding the numerical recorded average.

In which place value do the measurements begin to vary? ____________________
Both the *numerical average* and the recorded average will be rounded to that place.

Average all the measurements as they appear above. Round to the place value above.
This is the **numerical average**. ____________________

Next, round off each number in your data list to the place value above.
What is the **largest difference** between the **numerical average** and your **revised data**?
That number will be called the maximum likely error. ______________

Write the **recorded average** by using the numerical average ± the maximum likely error.

____

Verify your results with your instructor before proceeding.

Task 4: Using the method in task 3, find both the numerical and the recorded average of the measurements of a length (in cm) of rope listed below.

5.436, 5.429, 5.473, 5.492, 5.455, 5.415

Find the **numerical average**. ____________

Find the **recorded average.** ____________________

V. MAKING THE MEASUREMENTS

There are several limitations on a length measurement due to the material itself. For a piece of lumber finished at the mill by a machine driven planer, the measurements will be consistent over the surface to a very close tolerance. However, if the material is cut by a band saw that is guided by hand, then the variation in measurements of its dimensions at different locations may be quite large.

In this experiment you will make at least 5 measurements of length, width, and height using a variety of instruments to investigate the measurement of uniform or varying types of surfaces.

Task 5: Using the meter stick for measuring length in millimeters.

Before measuring your block of wood, examine the meter stick.
What is the smallest division on the meter stick?______________
How many significant digits can be recorded for a measurement up to the full length of the meter stick? __________________

When using a meter stick, it is not a good idea to measure to the edge of stick. Because this edge becomes worn with use, measurements from the edge may not be accurate. Start the measurement at 1 cm or 10 cm and then subtract the appropriate value from the reading to get the measurement. To get the most accurate reading, place the object next to the scale divisions of the meter stick and record to the smallest division.

DATA

Measure the length of your wood block in at least 5 different locations.

1		5	
2		6	
3		7	
4		8	

Find the **numerical average** length. __________________

Find the **recorded average** length. ______________________________

Task 6: Using the vernier calipers for measuring width in millimeters.

For instruction on how to use a vernier caliper refer to the Appendix A.

DATA

Measure the width of your wood block with the vernier caliper in at least 5 different locations.

1		5	
2		6	
3		7	
4		8	

Find the **numerical average** width. ____________________

Find the **recorded average** width. ______________________________

Task 7: Using the micrometer for making the height measurement in millimeters.

For instruction on how to use a micrometer refer to the Appendix B.

DATA

Measure the height with the micrometer in at least 5 different locations.

1		5	
2		6	
3		7	
4		8	

Find the **numerical average** height. ____________________

Find the **recorded average** height. ______________________________

VI. USING THE RECORDED AVERAGES TO CALCULATE OTHER VARIABLES

Because significant digits carry over when numbers are multiplied together, your choice for length, width, and height can make the volume calculated vary.

> When **multiplying** (or dividing) measurements together, round the calculated value to the same number of significant digits as the measurement with the **least number of significant digits** (the less accurate measurement).

For example: A rectangle has a length of 4.3 mm and a width of 2.72 mm.
Its area is 11.696 mm^2.
It must be rounded to 2 significant digits which is 12 mm^2.

It is necessary to use all the data for length, width, and height when finding both the numerical and recorded averages. The calculations can be tedious and are more easily done on a spreadsheet. The sorting process makes it easier to calculate the *smallest possible* volume and the *largest possible* volume.

Below are some guidelines for using the spreadsheet.
Put your data for each variable into its own column.
Sort, one by one, each column from the lowest to the highest value.
Into the next column, enter a formula for your calculation.
(The spreadsheet will do the calculations for you.)
Find the numerical average of the your calculation.
Find the recorded average.

Task 8: Finding the volume of the block using a spreadsheet.

What is the **equation** that you will use for finding the **volume** of your block, a rectangular solid? ______________________________

What type of **units** will be used for the volume? ____________

How many **significant figures** will you retain in the recorded result for the volume? ______

Find the **numerical average** of the volume. __________

Find the **recorded average** of the volume. ________________

Task 9: Finding the percent of error.

The *percent of error* will be the **maximum** of

$$\left|\frac{\text{the smallest average - the recorded average}}{\text{the recorded average}}\right| \text{ or the } \left|\frac{\text{the largest average - the recorded average}}{\text{the recorded average}}\right|.$$

Find the **percent of error** for the volume. ____________________

Next you will use a process similar to the one in task 8 to find the perimeter of the ***largest face*** of the block. Note that when measured values are added, the decimal positions are important.

When **adding** (or subtracting) measurements together, round the calculated value to the same decimal position as the measurement with the **least number of decimal positions** (the less precise number).

For example: One piece of wood is 25.1 cm long and another piece is 2.569 cm long.
The sum of the lengths is 27.669 cm.
It must be rounded off to 27.7 cm.

Task 10: Finding the perimeter of the largest face of the block using a spreadsheet.

What type of **units** will be used for the perimeter? ____________

How many **significant figures** will be recorded in the result for the **perimeter?**

______

Find the **numerical average** of the perimeter. ____________

Find the **recorded average** of the perimeter. ____________________

Find the **percent of error** for the perimeter. ____________________

All averages on pages 6 - 9 should have the unit of measure included.

VII. SUMMARY

This laboratory has been written to set the tone for how you will collect and report data in future investigations. It is meant to show you not only the importance of making careful measurements but more importantly how to make mathematically sound decisions for reporting your calculated results.

Keep this laboratory handy and use it as a reference guide as you continue your studies.

Task 11: Summarizing Engineering Measurement.

Using complete sentences, write a paragraph or two describing how you will collect data in the future and how you will make decisions on rounding off your computed results.

Appendix A: THE VERNIER CALIPER

A vernier caliper measures just like any other ruler when using the main scale. Where the marker lines up for the tens, units, and first decimal position, indicates the measurement for these values. Then the vernier scale is used to find the next decimal position.

Referring to the figure below, the scale is closed. In this position the jaws are together indicating a measurement of zero. The zero on the main scale lines up with the zero on the vernier scale indicating a scale reading (measurement) of zero. The Vernier scale has 10 divisions that cover only 9 divisions on the main scale. This means that each vernier division is 1/10 smaller than the corresponding main scale division. Only one vernier scale division will line up with a main scale division for any position of the vernier scale.

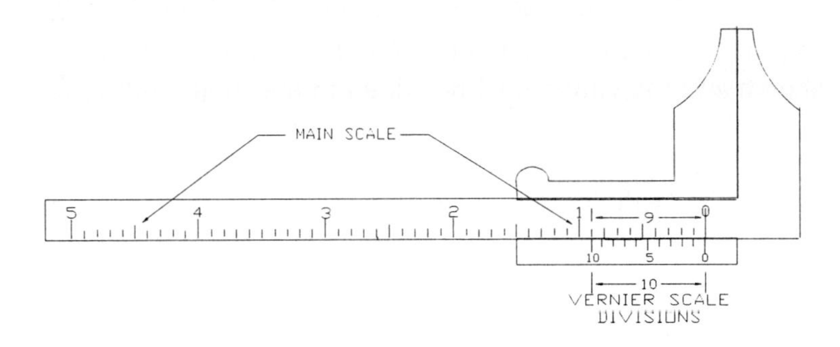

In the figure below a measurement has been made. The jaws have opened indicating a measurement. The 0 points on the main and vernier scale have moved apart the same amount indicating the same measurement. This measurement is indicated as 4.1 cm. assuming that the vernier is calibrated in metric units. On the vernier scale the 7th line is the only line that lines up with a line on the main scale. Therefore this indicates a vernier measurement of .07 which must be added to the reading for the main scale. The final reading then is 4.17 cm.

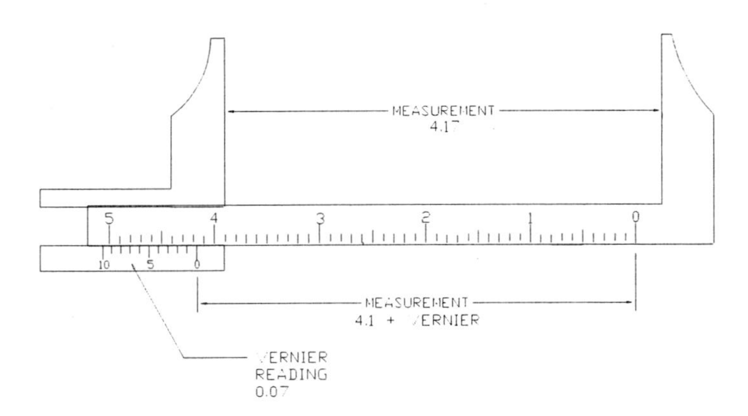

Appendix B: THE MICROMETER

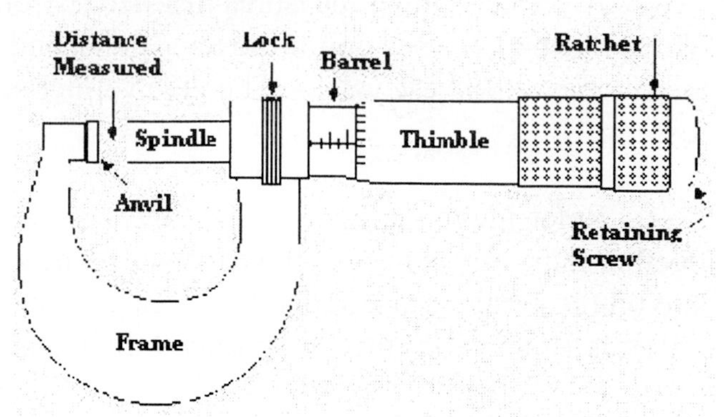

Refer to the diagram on the right to learn the names of the key parts of a micrometer. This description uses a micrometer with a 25 millimeter range (1 inch).

When using a micrometer, first check its zero reading. With the spindle closed, the micrometer should read 0.00. The zero on the thimble lines up with the zero on the barrel of the micrometer. The edge of the thimble should line up with the first vertical line on the fixed scale on the barrel. If there is a reading above 0 on the thimble, then this indicates a measurement at a zero reading. This value must be subtracted from the final reading to get the measurement. If the thimble reads below zero, then a reading below zero is shown and the value must be added to the final reading to get the measurement.

Each complete rotation of the thimble represents 1/2 of a millimeter (0.02 inch) movement of the spindle. For each rotation of the thimble, the edge will line up with a vertical line on the barrel, and after 10 rotations of the thimble the number 5 appears. This represents a measurement of 5 millimeters.

The markings on the thimble represent a range of 0.00 to 0.50 of a millimeter. The readings off the thimble must be added to the readings from the barrel when recording each measurement.

For example: If the barrel reads 6.0 mm and the thimble reads 0.35mm, then the measurement is 6.35mm.

If the barrel reads 7.5 mm and the thimble reads 0.14 mm, then the measurement is 7.64 mm.

MATHEMATICS LABORATORY INVESTIGATION

GEODESIC DOMES

Topic: **SOLUTION OF TRIANGLES**
Prerequisite knowledge: *Law of Cosines, Right Triangle Trigonometry*

I. INTRODUCTION:

Geodesic domes are large, geometric structures that are used to enclose space. One of the most visible geodesic domes is the "Spaceship Earth" at Epcot Center in Florida.

Geodesic domes were studied and developed by the famous scientist, mathematician and architect: R. Buckminster Fuller. Along with thousands of other inventions, Fuller invented domes for use as houses, buildings and various devices such as radar antennae.

Domes are often used as approximations of spheres. Standard building materials and labor costs are just two of the factors that can make it unfeasible to use actual spheres as design elements.

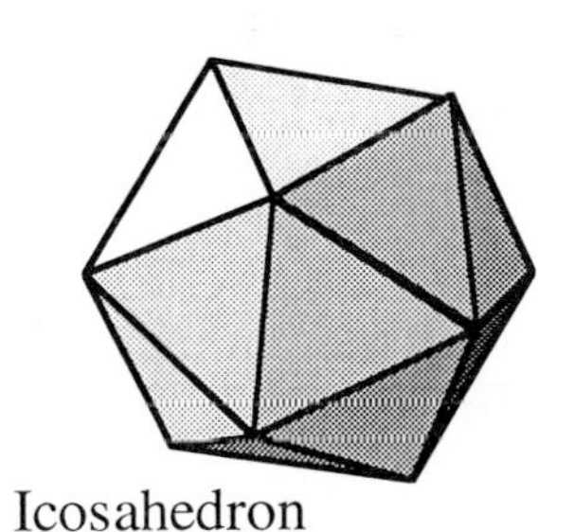

Icosahedron

This investigation focuses on domes that are based on the solid known as the *icosahedron*. An icosahedron is regular polyhedron with 20 congruent *faces*, each of which is an equilateral triangle; five faces meet at each *vertex*, or corner of the solid.

While the icosahedron is probably not what you picture when you think of a dome, it is the underlying structure for many domes. By dividing each face into smaller triangles and then "bumping out the vertices" we can create a dome that still has triangular faces but is much closer to the shape of a sphere. (See the photo below.)

The first step in this process will be to divide each strut in half. This sub-division is called a *two-frequency sub-division (denoted 2V),* since it has broken all of the triangles' edges into two parts. A *matrix* connecting the midpoints of the edges is drawn on each of the triangular faces and finally each of the intersection points (called *nodes*) in the matrix is pushed out to create a vertex on the 2V dome. The distance between the center of the dome and any vertex is equal to the radius of the dome.

A three-frequency (3V) sub-division, likewise, breaks the original struts into three equal parts, forming a matrix with nine small triangles on each face of the icosahedron. The nodes are again pushed out so that the distance between them and the center of the dome is equal to the radius of the dome. The more sub-divisions of the triangles of the Icosahedron that are made, the closer the shape gets to a sphere.

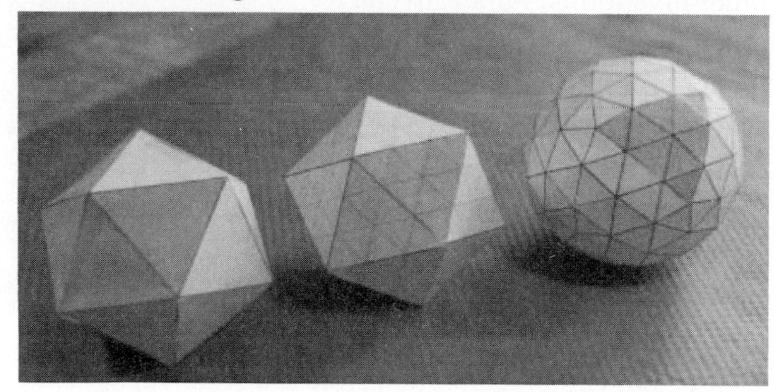

II. PREPARATION

Each group requires three models: one with a 2V matrix on its faces, one with the 3V matrix (net diagrams are attached for each of these) and a wire frame model (directions are attached).

This investigation relies heavily on your knowledge of the Law of Cosines. Review this concept paying special attention to the requirements for using it. What parts of a triangle do you need to know in order to apply the Law of Cosines?

Throughout this investigation, do your calculations to the nearest 0.001cm and 0.1^0.

III. WORKING IN THREE DIMENSIONS

One of the most challenging aspects of working in 3 dimensions is finding an appropriate plane in which to define a problem or a particular part of a problem. As you work, you will constantly be using the 3-dimensional models to assist you in locating specific triangles and then sketching the triangles in 2 dimensions in order to use your knowledge of trigonometry to solve for various parts of the triangle. You will go back and forth between the 3 dimensional model and various 2 dimensional sketches.

You may get frustrated at times, but remember to use the 3D models to help you to make decisions on the relative sizes of lines and angles.

IV. THE 2 FREQUENCY (2V) ICOSAHEDRAL DOME

TASK 1: The icosahedron.

When you are taking measurements on the wire frame model remember that the strut length is the distance between two vertices, not the length of the material used. It is standard practice to take the measurements several times and use the average.

a) Measure and record the radius of the icosahedron to at least the nearest 0.1 cm. (It may be easier to measure the diameter.)

radius (r) = __________

b) Measure and record the length of a strut to at least the nearest 0.1 cm. Be as accurate as possible.

strut length = _________

Are all of the struts of equal length? __________

c) Consider a triangle composed of 2 radii and a strut. In the sketch below, label the sides with their measures, calculate and label the measure of the central angle and the axial angles.

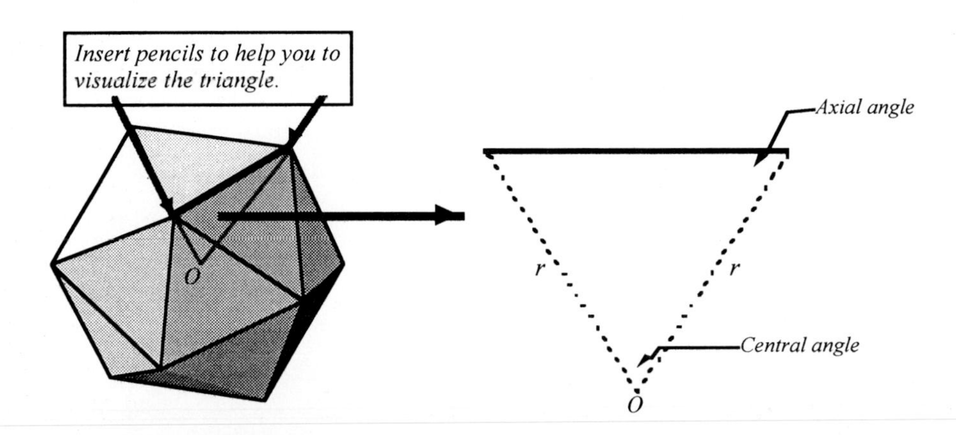

TASK 2: <u>The 2 frequency matrix.</u>

Use the model with the 2V matrix to help you to visualize the location of A and B on the wire frame model.

a) <u>On the surface</u> of the icosahedron,

What is the length of A? ____________

What is the length of B? ____________

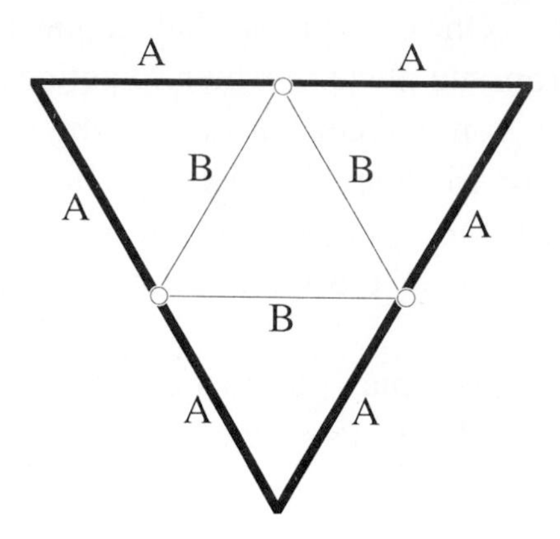

Each node (denoted with an open circle) will be pushed out radially to become a vertex of the 2V dome. When the nodes are pushed out, how will the lengths of A and B be affected? Will they get longer, shorter, or stay the same?

TASK 3: <u>Finding the new length for strut A'.</u>

Refer to the triangle that you labeled in Task 1 to help you to:

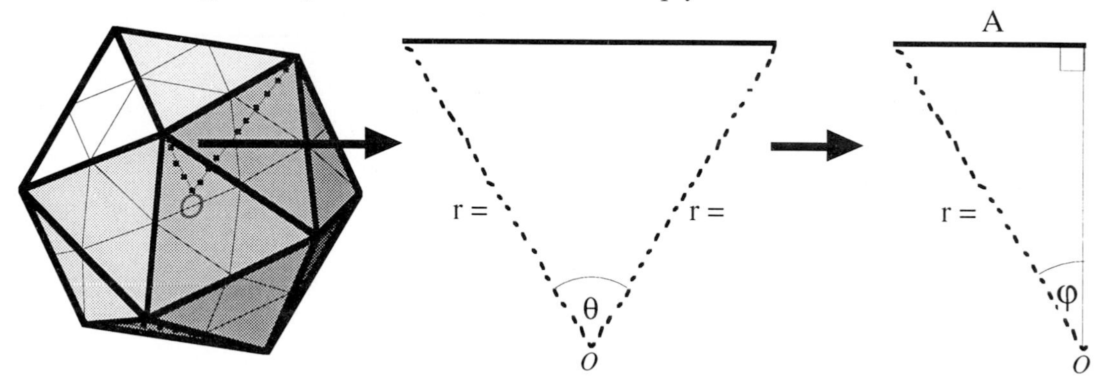

a) Fill in the values for A and r.

b) What is the measure of angle θ? ________ of angle φ? _________

c) Now "bump out" the node. Remember that the distance between a vertex and the center of the dome is equal to the radius. Calculate the length of the new strut A'. Record your answer here.

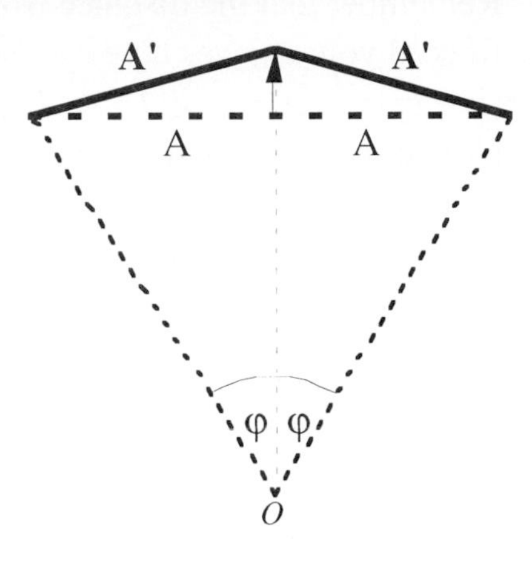

TASK 4: <u>Finding new strut length B'.</u>
(Get ready for some mind twisting activity. Remember to use your 3D models to help you with these questions.)

We now know the length of new strut A'. What about the other 3 new struts? How do you think their lengths will compare to the length of new strut A'?

a) Consider the triangle on the original icosahedron composed of B and the lines connecting the endpoints of B to the center. How long are the sides of this triangle. (Hint: This length appeared in Task 3. It is not the radius of the original icosahedron. Use your knowledge of right triangular trigonometry to calculate this length.) Record your answer here: _____________

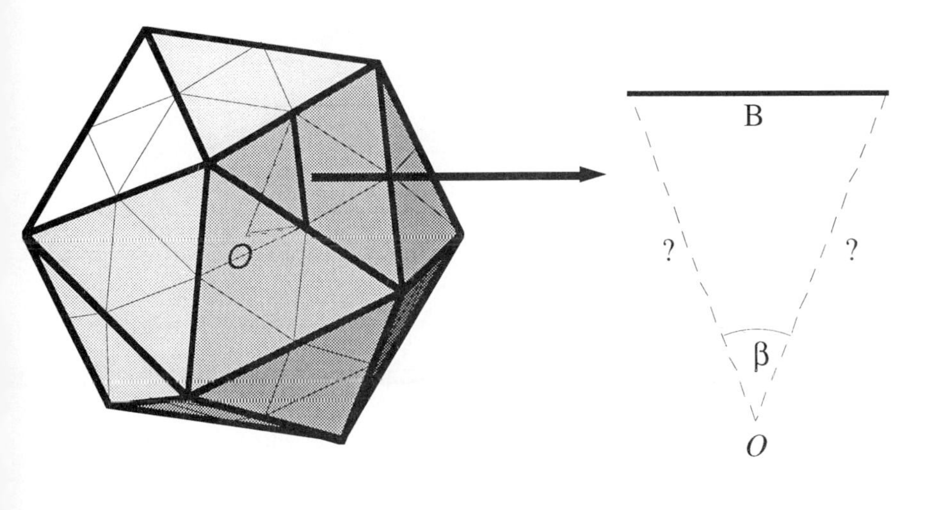

b) What is the length of B?

____

c) Calculate the measure of the angle labeled β.

____

d) Now "bump out" the nodes to create vertices and calculate the length of new strut B'. Remember that the distance from a vertex to the center is always equal to the radius. Record your answer here: __________

Notice that β stays the same.

V. SUMMARY OF THE 2V ICOSAHEDRAL DOME

Fill in your results for the 2V icosahedral dome:

radius = _______ length of strut A' = ________ length of strut B' = ________

How many struts of each type would be necessary to build a complete dome?

# of struts A' = _______ # of struts B' = ________

Fill in the measures of A' and B' on the illustration on the right.

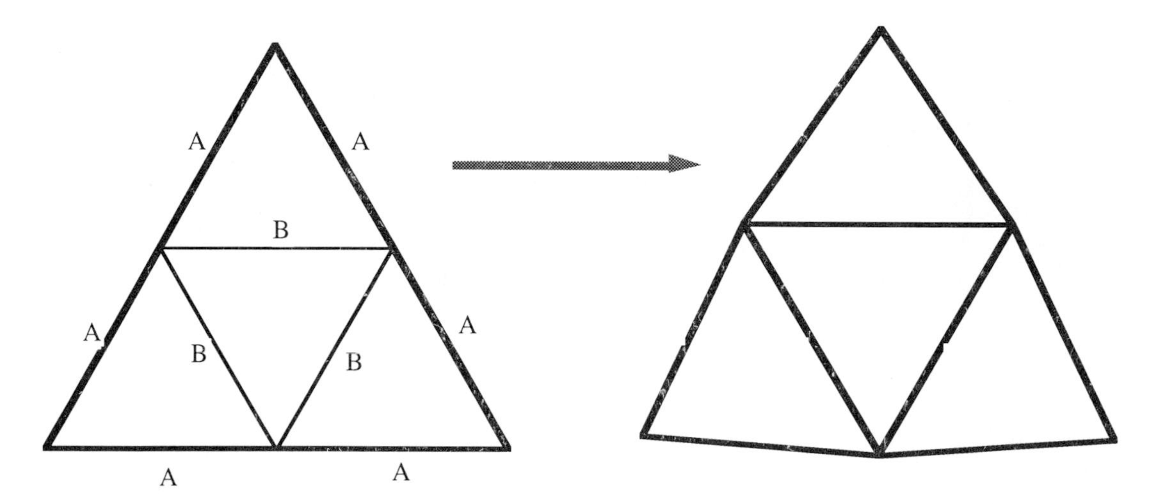

Fig. 1 One face of an icosahedron with a 2V matrix.

Fig. 2 The corresponding portion of a 2V icosahedral dome.

If you make 20 copies of Fig. 2 you can construct a model of a 2V icosahedral dome.

Reprinted with permission from Bowler's Journal International.

The National Bowling Stadium dominates the skyline in Reno, Nevada. Its focal point in is an icosahedral dome which houses a 100-seat theater.

VI. THE 3 FREQUENCY (3V) ICOSAHEDRAL DOME

TASK 5: The 3 frequency matrix.

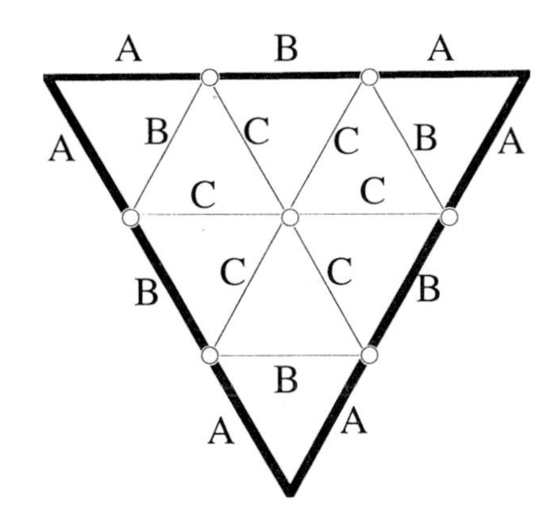

Use the model with the 3V matrix to help you to answer the following questions.

a) On the surface of the icosahedron,

What is the length of A? ____________

What is the length of B? ____________

What is the length of C? ____________

b) Each node (denoted with an open circle) will be pushed out to become a vertex of the 3V dome. When the nodes are pushed out, how will the lengths of A', B' and C' compare to the lengths of A, B, and C? Will they be longer, shorter, or stay the same?

TASK 6: <u>Finding the length for struts A' and B'.</u>

Consider the triangle composed of one of the original struts of the icosahedron and two radii.

a) Label the sketch of this triangle with all of the measurements that you already know.

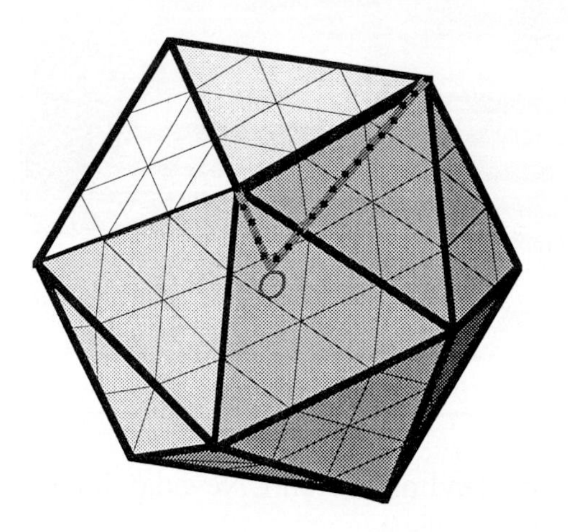

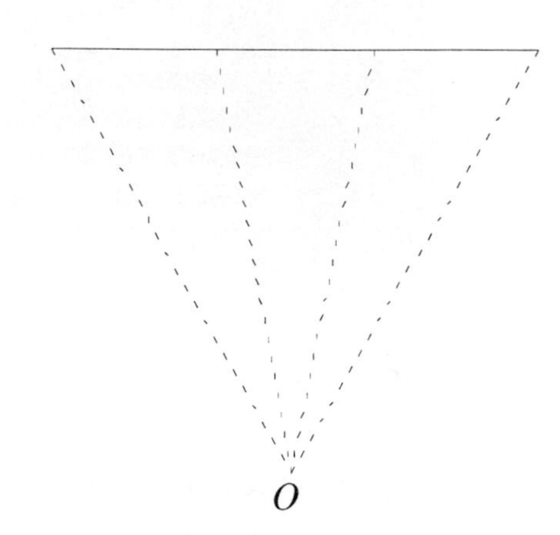

b) Use your knowledge of trigonometry to calculate the value of the unknown segments and angles.

c) Now "bump out" the nodes. Find the lengths of the new struts A' and B'. Record your answers here:

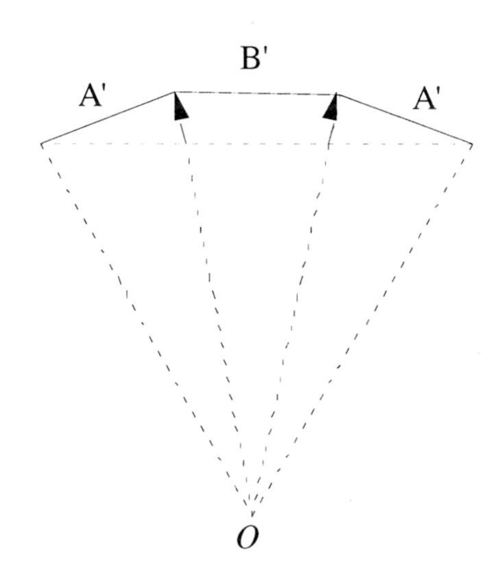

TASK 7: Find the length of strut C'.

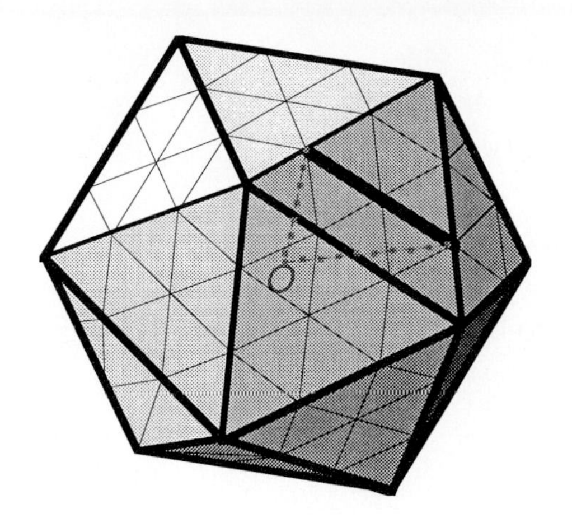

a) Sketch the triangle shown in the 3D diagram. Label the lengths of all three sides of the triangle. (Be careful! This is not easy.)

b) Using methods similar to the other sections of the investigation, find the length of strut C'. (Hint: Remember that the distance between any vertex and the center is equal to the radius.) Record your answer here.

VII. SUMMARY OF 3V ICOSAHEDRAL DOME

Fill in your results for the 3V icosahedral dome:

radius = ______ length of strut A'= _________

length of strut B' = _________ length of strut C' = _________

How many struts of each type would be necessary to build a complete dome?

# of struts A' = _______ # of struts B' = _______ # of struts C' = _______

Fill in the measures of A', B' and C' on the illustration on the right.

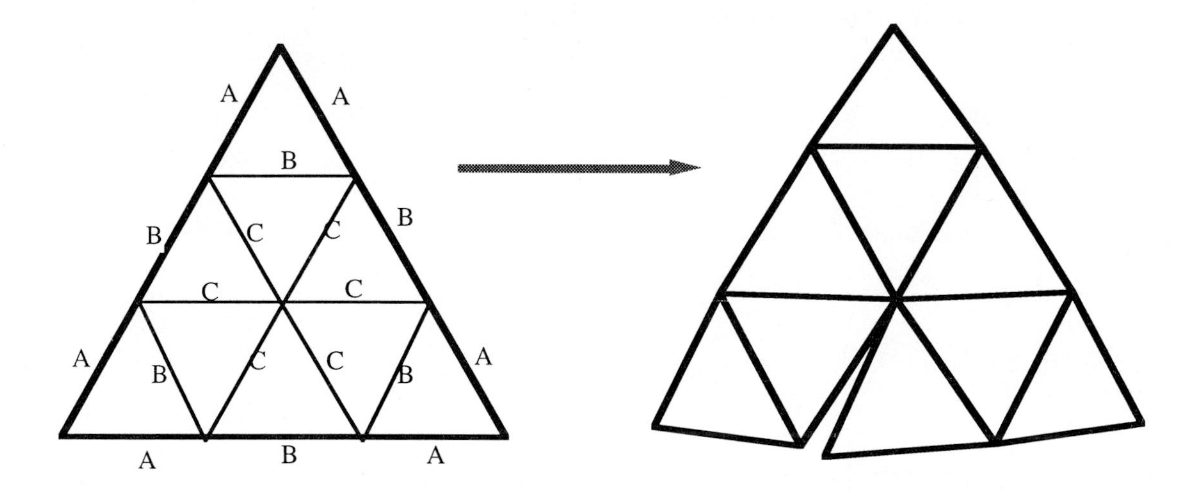

Fig. 1 One face of an icosahedron with a 3V matrix.

Fig. 2 The corresponding portion of a 3V icosahedral dome.

If you make 20 copies of Fig. 2 you can construct a model of a 3V icosahedral dome.

VIII. EXTENSIONS

1) Build models of the 2V and 3V icosahedral domes.

   All of the mathematics that you just completed can be generalized and used to build 2V and 3V icosahedral domes of any size. It's simply a matter of proportion. The scale factor for each strut is called the chord factor. It is a number which when multiplied by the radius gives the strut length for a dome: Strut Length = (Chord Factor)(Radius)

   Find the chord factors for both the 2V and 3V domes. (Hint: Consider the radius to be equal to 1 unit.)

2) After you build the model of the icosahedron, notice that there are five triangles touching at each vertex. Investigate the 2V and 3V icosahedral domes. Are there always five triangles touching at a vertex?

3) A dome which will sit flat on the ground can be made by removing the five triangles on the bottom of an icosahedron. Can the 2V and 3V icosahedral domes be "cut off" so that they will sit flat? How? What will the heights of these domes be in comparison with their diameters?

   Note: The dome pictured on the first page of this investigation is located at Epcot in Walt Disney World in Florida. It is the first completely spherical geodesic dome ever built. Previous geodesic dome structures were built so that they sat flat on a horizontal surface. "Spaceship Earth" stands 15 feet above the ground and is supported by three pairs of steel legs.

4) A net diagram is the pattern that a geometric solid forms as it is unfolded onto a two-dimensional surface. Create net diagrams for the 2V and 3V domes. Can you create a single net for each of the domes that just needs to be cut out, folded and taped or do you have to split the nets into two or more sections? See if you can minimize the amount of tape necessary.

5) Investigate domes built on the octahedron and tetrahedron.

GLOSSARY

Axial Angle An angle formed by a strut and a radius at a vertex.

Central Angle The angle created by two radii that contain the endpoints of a strut.

Center (*O*) The center of the sphere that circumscribes the dome.

Frequency(*V*) A number used to describe the complexity of the dome. It is the number of struts that replace a single strut in the icosahedron. (See examples below.)

Matrix The triangular grid that is drawn on the faces of the icosahedron in preparation for increasing the frequency of a dome.

Node A point of intersection on the matrix.

Radius (r) The distance from the center to a vertex.

Strut A line segment that creates an edge of the dome. (An icosahedron has struts.)

Vertex A point of intersection of two or more struts. Vertices lie on the sphere that circumscribes the dome. (An icosahedron has 12 vertices.)

1-frequency (1V) | 2-frequency (2V) | 3-frequency (3V)

APPENDIX - BUILDING THE MODELS

Each group requires three models: an icosahedron with a 2-frequency matrix, an icosahedron with a 3-frequency matrix, and a wire frame of an icosahedron. The net diagrams for the icoashedrons that follow have been sized to fit on a single page. They are too small for the actual models. Either enlarge them so that they will approximately match the wire frame model or have the students create them in an appropriate size. Students can draw one equilateral triangle to size and then use it as a template. Using card stock will make models that are fairly sturdy.

If the model building is assigned as a project you may get some models that are strong enough to be used repeatedly.

Commercially produced kits for building platonic solids are available.

Wire Frame Model

Wire frame models can be made from coffee stirrers, very thin straws 4.5"-5" long, and pipe cleaners. Use a whole coffee stirrer for each strut and join them together with pipe cleaners at the vertices. By cutting the pipe cleaners to lengths that are the same as the stirrers the models will be very stable and will not fall apart easily. If you cut the pipe cleaners too short the models will continually pop apart - very frustrating!

You may be surprised at how many pipe cleaners fit into a single coffee stirrer. Since five struts will be joined at each vertex you will need to "stuff" more than one pipe cleaner into some straws. That will have the effect of making the models straonger.

There are many other ways of making these models: toothpicks and miniature marshmallows, bamboo skewers and gumdrops, stiff wire, straws strung together with string,... Your students may be very creative, especially if the model building is a project.

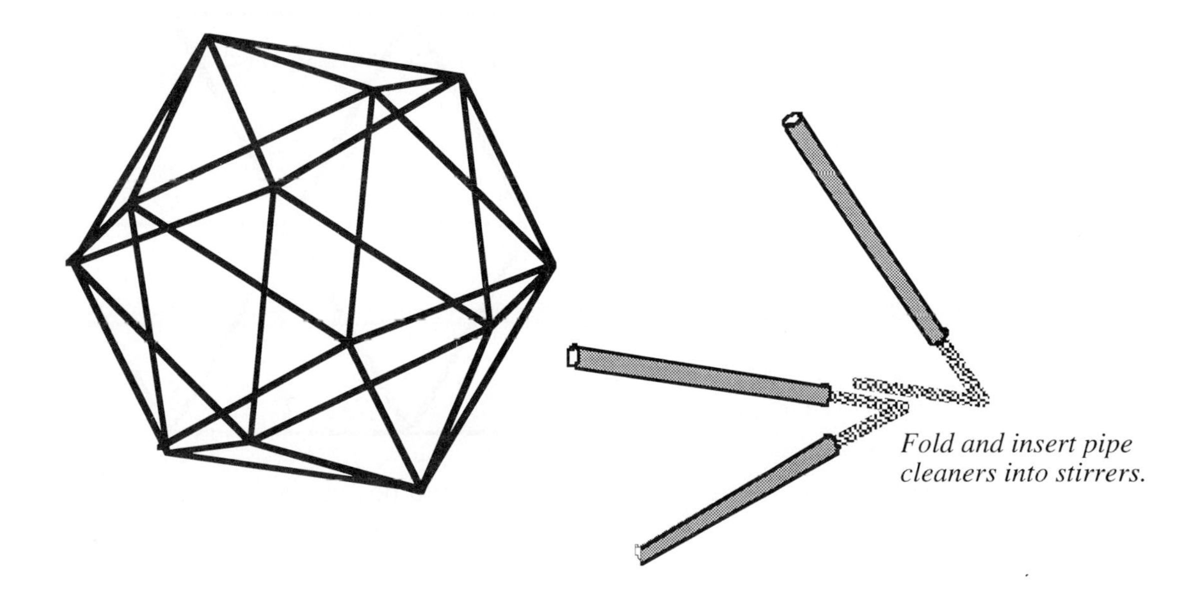
Fold and insert pipe
cleaners into stirrers.

Net Diagram for an Icosahedron with No Matrix

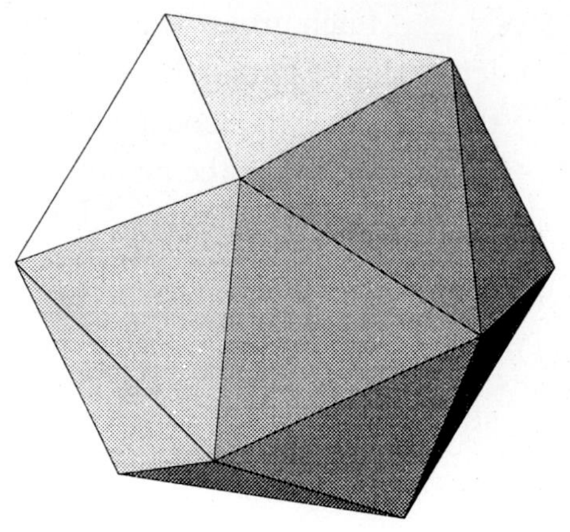

1) Cut along the perimeter of the net diagram.

2) Fold on the solid lines. (Easily done if you first draw over them with a ball point pen using enough pressure to slightly score the solid lines.)

3) Tape to form an icosahedron.

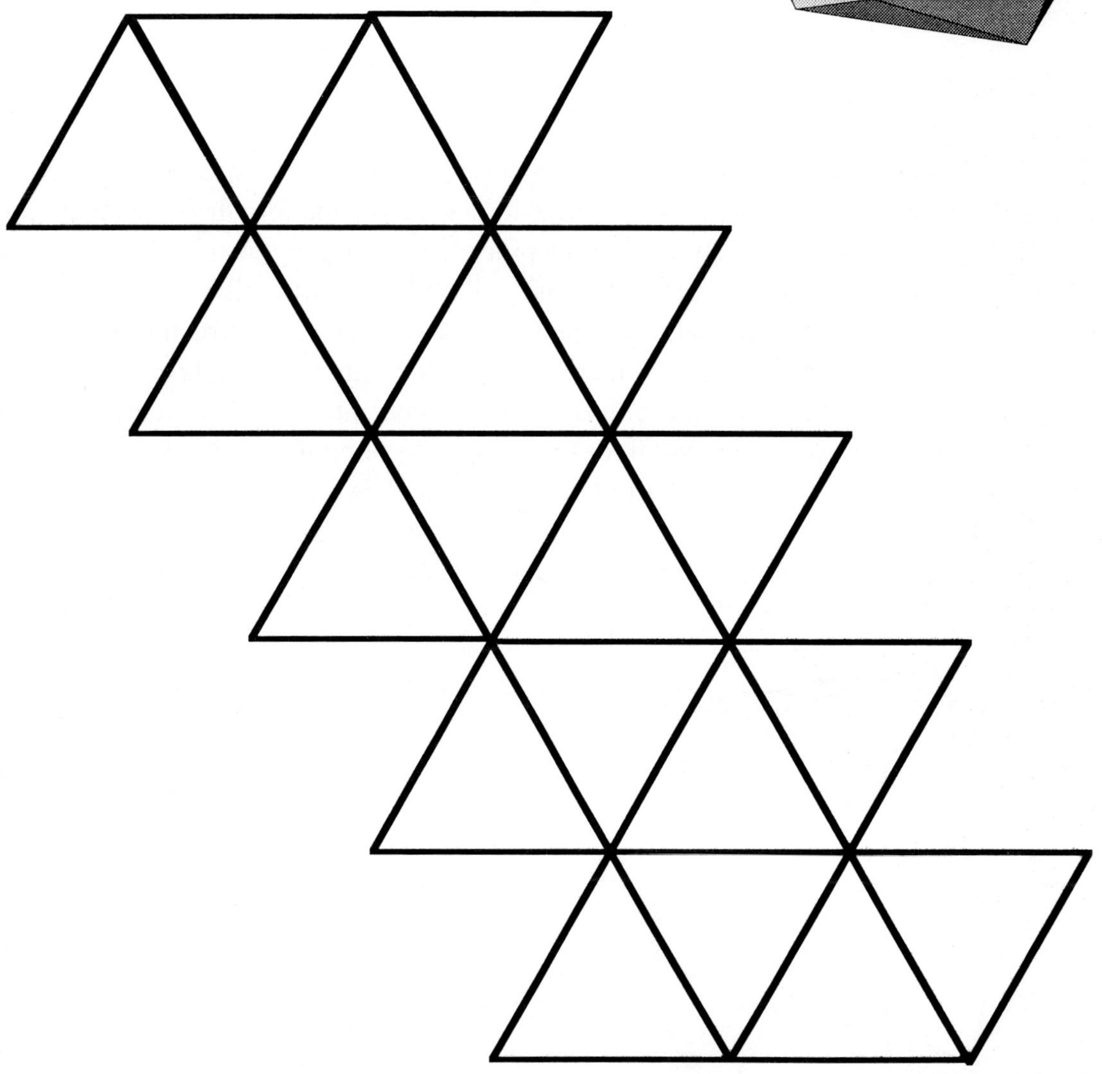

This net diagram was sized to fit on a single page. It should be enlarged to make a useable model.

Net Diagram for an Icosahedron with a two-frequency (2V) Matrix

1) Cut along the perimeter of the net diagram.

2) Fold on the thick lines. (Easily done if you first draw over them with a ball point pen using enough pressure to slightly score the thick lines.)

3) Tape to form an icosahedron.

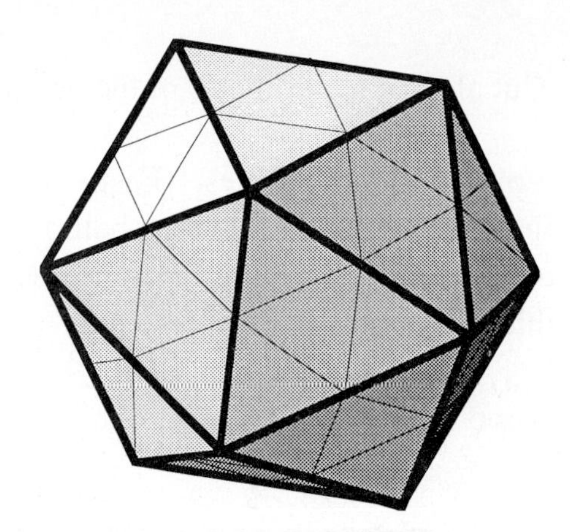

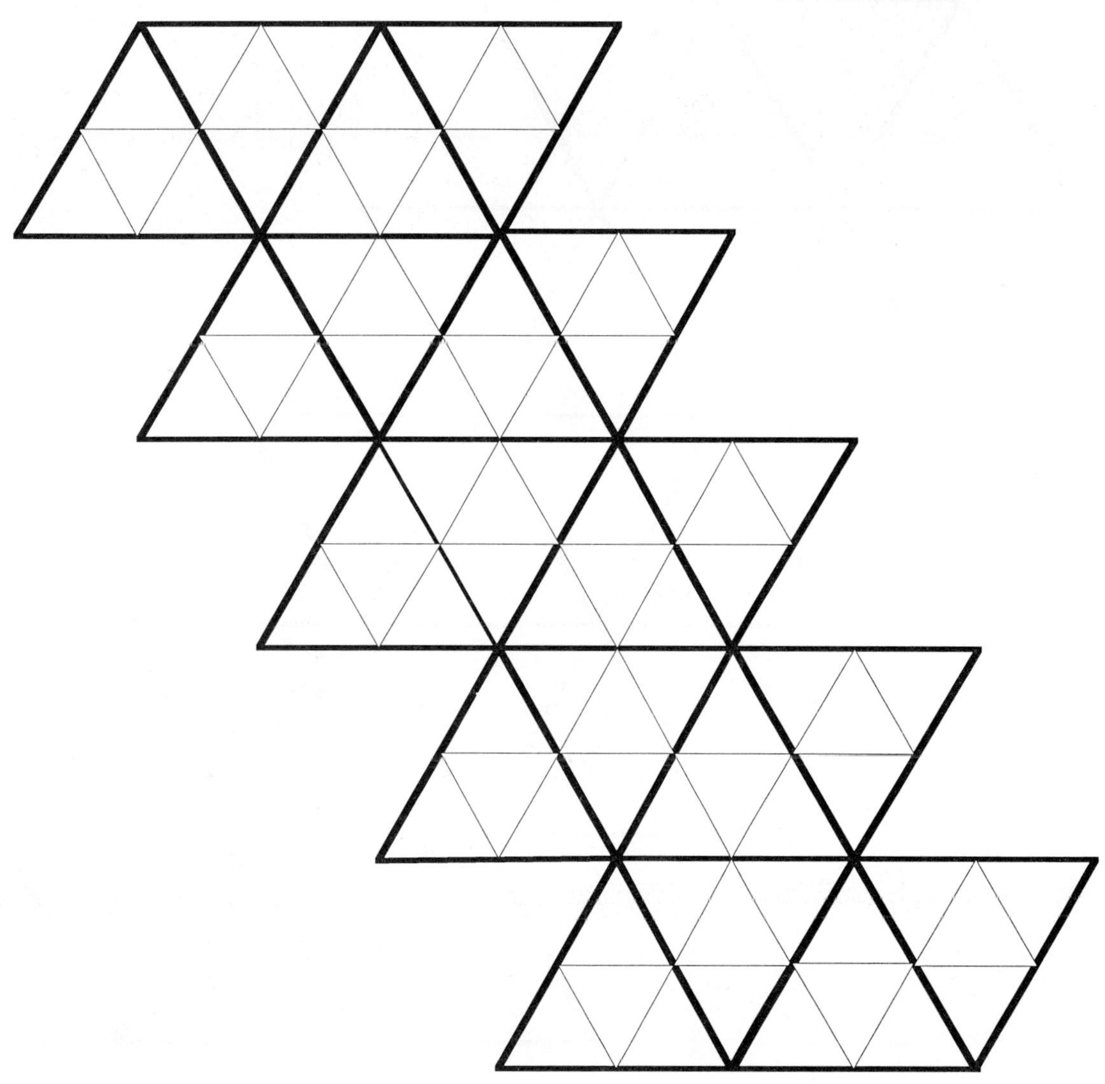

This net diagram was sized to fit on a single page. It should be enlarged to make a useable model.

Net Diagram for an Icosahedron with a three-frequency (3V) Matrix

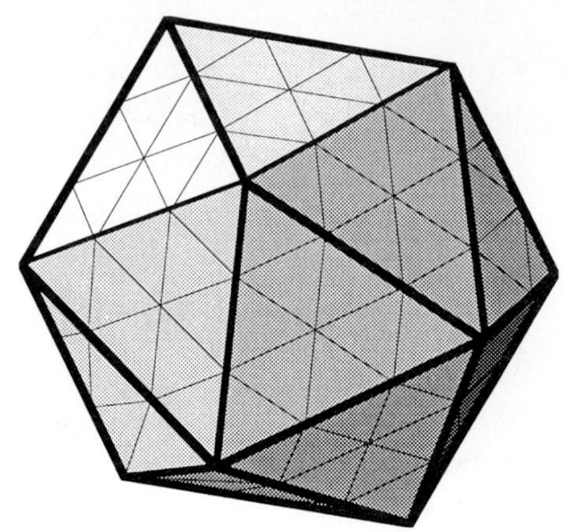

1) Cut along the perimeter of the net diagram.

2) Fold on the thick lines. (Easily done if you first draw over them with a ball point pen using enough pressure to slightly score the thick lines.)

3) Tape to form an icosahedron.

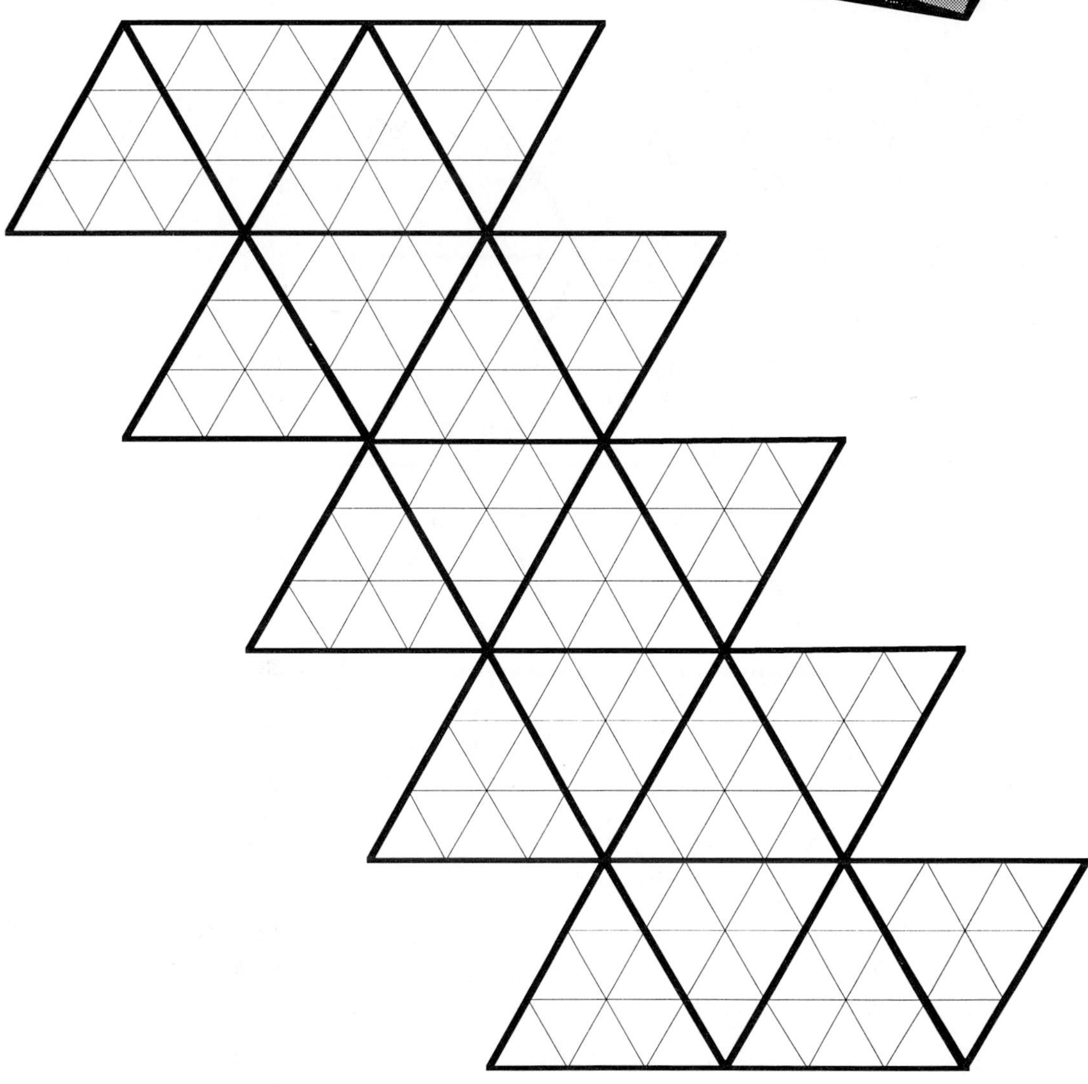

This net diagram was sized to fit on a single page. It should be enlarged to make a useable model.

Alternate to Net Diagrams

Assemble so that five triangles meet at each vertex.

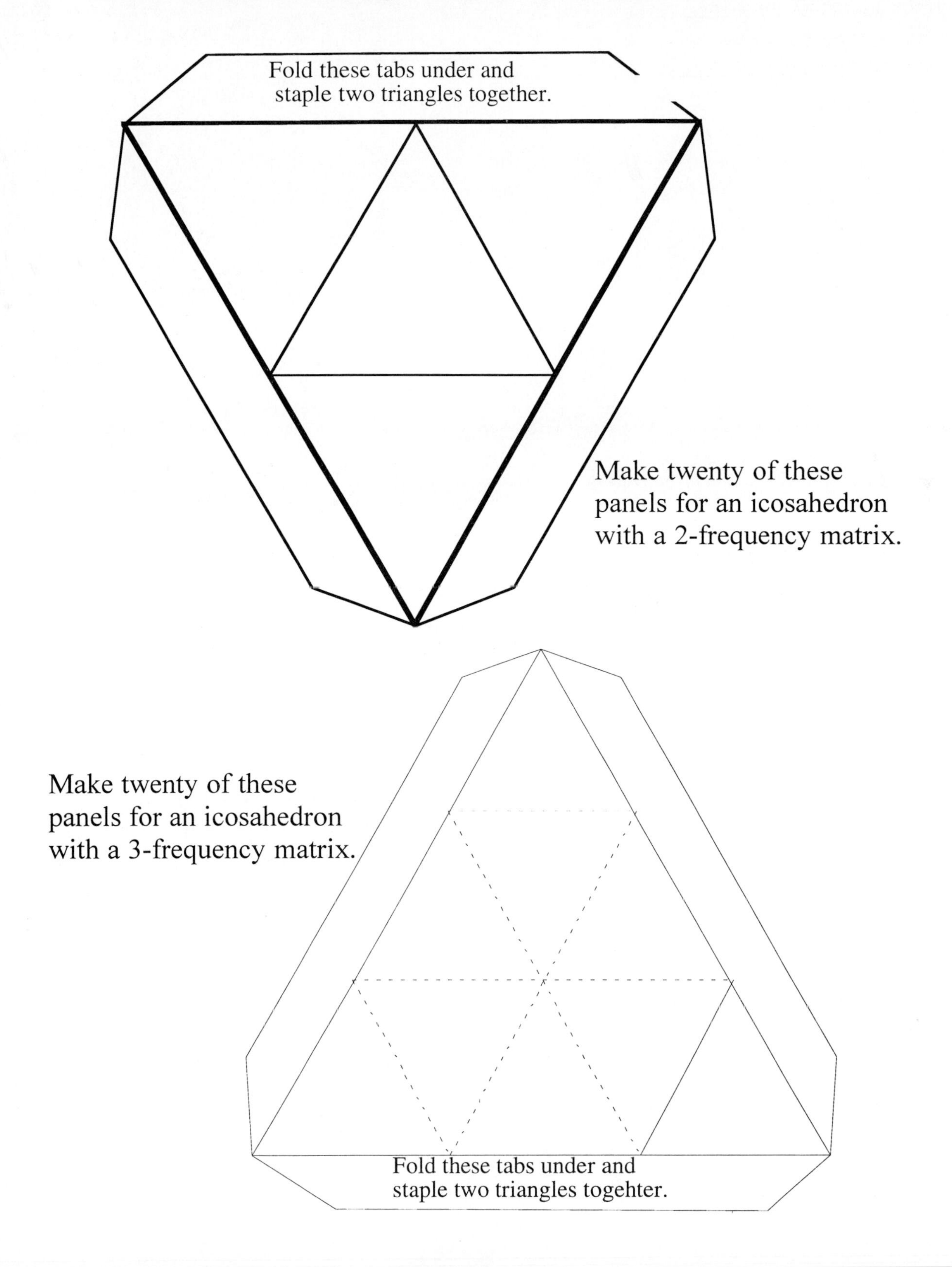
Fold these tabs under and
staple two triangles together.
Make twenty of these
panels for an icosahedron
with a 2-frequency matrix.
Make twenty of these
panels for an icosahedron
with a 3-frequency matrix.
Fold these tabs under and
staple two triangles togehter.

MATHEMATICS LABORATORY INVESTIGATION

GEOMETRY OF DESIGN - PROPORTIONING SYSTEMS

Topic: **SLOPE, INTRODUCTION TO CONCEPT OF LIMITS**
Prerequisite knowledge/skills: *Use of spreadsheet to enter data via formulas (or via sequences if using TI-82), solving quadratic equations, calculating slope, similar rectangles*

I. INTRODUCTION:

"...architecture ... exists in any building, great or small, whose geometry creates a mathematical proportion." LeCorbusier

Artists in many fields have searched for a definition of beauty, for a formula which will produce good design. Paintings, sculpture, buildings, music, poetry all have been studied and analyzed in detail in the hope that the secrets of their creation would be revealed. The quest has continued from Pythagoras (500 BC) who thought that certain numerical relationships were the basis of the universe, to the development of the Ken which orders the geometry of Japanese architecture, to the Renaissance artists with their geometrical rules for design, to current attempts to create intelligent architecture machines.

Even though such a formula doesn't seem to exist, the search has uncovered some interesting theories. In particular several proportioning systems have become accepted as ways of providing unifying relationships among the parts of a design. These systems create a sense of order and harmony which seem to be perceived by the observer of the design.

A study of proportioning systems leads to a study of number patterns. This investigation will consider two systems; one is based on *geometric* sequences of numbers and the other on *additive* sequences of numbers.

II. DEFINITIONS

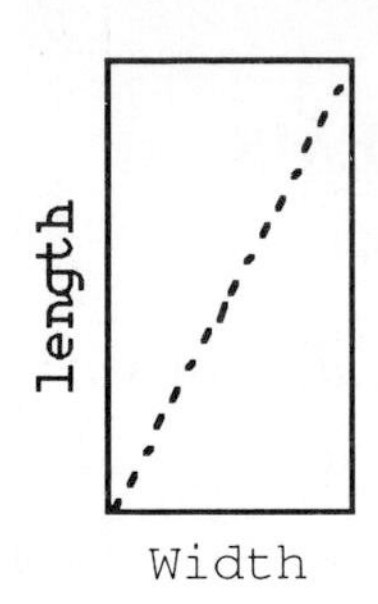

For the purposes of this investigation, we will consider rectangles that are not squares. We will call the longer side the length and the shorter side the width. Generally the rectangles will be positioned so that the sides are horizontal (width) and vertical (length).

Consider a diagonal drawn from the lower left corner to the upper right corner of a rectangle. How would you calculate the slope of the diagonal? ____

III. GEOMETRIC SEQUENCES

A *geometric sequence* is a list of numbers in which each term is a constant multiple of the one before. For example, 1, 3, 9, 27, 81, 243,... is a geometric sequence for which the constant multiple is 3.

In order to apply this sequence to design, use adjacent terms to create a sequence of rectangles:

rectangle 1 1 x 3
rectangle 2 3 x 9
rectangle 3 9 x 27
rectangle 4 27 x 81
⋮ ⋮

TASK 1: Creating examples based on geometric sequences.

Create three of your own examples of geometric sequences:

1)____ (constant multiple is _____)

2)____ (constant multiple is _____)

3)____ (constant multiple is _____)

Set up a spreadsheet to list the widths, the lengths and the slopes of the diagonals of 10 rectangles produced by one of your geometric sequences.

Rectangles based on a Geometric Sequence

	Width	Length	Slope of Diagonal
rectangle 1			
rectangle 2			
rectangle 3			
rectangle 4			
rectangle 5			
rectangle 6			
rectangle 7			
rectangle 8			
rectangle 9			
rectangle 10			

This sequence of slopes is an example of a *constant* sequence because all of the slopes are the same (or are constant). Why did the slopes all turn out to be the same number?

TASK 2: Finding the *regulating lines.*

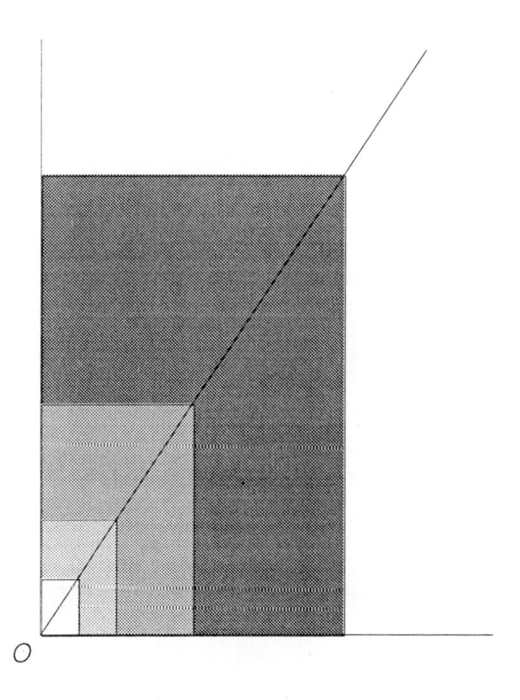

If you draw the rectangles in your sequence so that the left bottom corner of each one is placed at the origin of a Cartesian coordinate system, and then draw in the diagonal of each rectangle from the origin to the upper right corner, you will see that all of the diagonals lie on the same line.

This diagonal is called a *regulating line* and is used by artists to create similar rectangles.

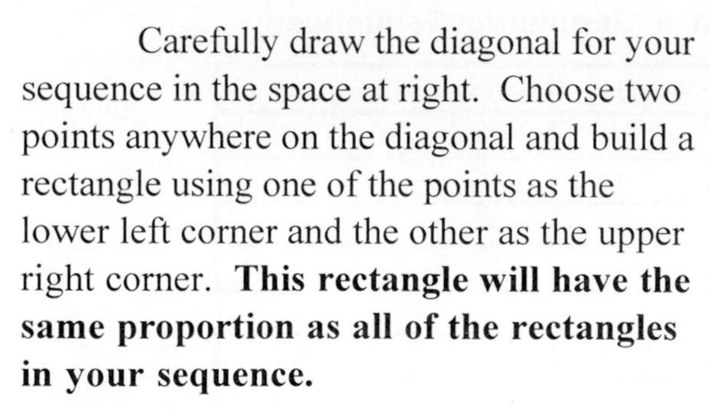

Carefully draw the diagonal for your sequence in the space at right. Choose two points anywhere on the diagonal and build a rectangle using one of the points as the lower left corner and the other as the upper right corner. **This rectangle will have the same proportion as all of the rectangles in your sequence.**

Why?

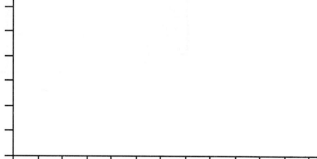

Draw a line perpendicular to the diagonal line and choose two points on it. Build a rectangle using those two points as the upper left corner and the lower right corner. What is the proportion of this new rectangle?

How does it relate to the proportion of the original rectangles?

IV. PROPORTIONING SYSTEM BASED ON SIMILAR RECTANGLES

TASK 3: Locating similar rectangles in the Ca D'Oro (drawing attached).

Trace the rectangle of the entire building as outlined on the drawing on a piece of tracing paper. Draw a regulating line from the lower left corner of the rectangle to the upper right corner. Move the tracing paper around the drawing and line up the lower left corner of your regulating diagram with other parts of the drawing. See how many similar rectangles you can find. (Don't limit yourself to a certain size or direction.) Highlight the rectangles that you find on the drawing.

Since this is a drawing and since it has been reproduced many times, some of the rectangles in may not be exact. There are also decorations and embellishments which may not have been part of the original design. Allow for a little variation as you work from the drawings in this lab.

V. ADDITIVE SEQUENCES

1, 3, 4, 7, 11, 18, ... is an example of an *additive sequence*. Notice that each term is the sum of the two terms preceding it.

TASK 4: Proportioning based on additive sequences.

Create a different additive sequence for every person in your group.

Name:____________________ Sequence________________________________

Name:____________________ Sequence________________________________

Name:____________________ Sequence________________________________

Name:____________________ Sequence________________________________

Individually, create a sequence of rectangles as in part III using adjacent terms for the widths (horizontal) and lengths (vertical).

Also individually, create a spreadsheet like the one below to list the widths, lengths, and slopes of the diagonals of 15 rectangles produced by your additive sequence.

Rectangles based on an Additive Sequence

	Width	Length	Slope of Diagonal
rectangle 1			
rectangle 2			
rectangle 3			
rectangle 4			
rectangle 5			
rectangle 6			
rectangle 7			
rectangle 8			
rectangle 9			
rectangle 10			
rectangle 11			
rectangle 12			
rectangle 13			
rectangle 14			
rectangle 15			

Compare your results. Discuss them here. (You may want to compare with other groups as well.)

The sequence of slopes for these rectangles is a *convergent* sequence because the terms get closer and closer to a particular value. To what number did the sequence of slopes *converge*? ________________ This number is called the *limit* of the sequence.

On a sheet of graph paper, draw as many of the rectangles in one of your groups' additive sequence as you can. Be sure to locate the lower left corner of each rectangle at the same point. Draw in the diagonals which connect the lower left corner of each rectangle to the upper right corner of the rectangle. What do you notice about the diagonals?

The further along a rectangle is in an additive sequence, the closer it is to the rectangle which is called the golden rectangle and which plays an important part in art, architecture and in many other fields.

TASK 5: Constructing a golden rectangle.

Draw a square (carefully!). Find the midpoint of one of the sides and connect it to one of the vertices on the side opposite. Using a compass, draw a circular arc from that vertex of the square. The point at which the arc meets the extension of the side of the square will become the corner of a golden rectangle as shown.

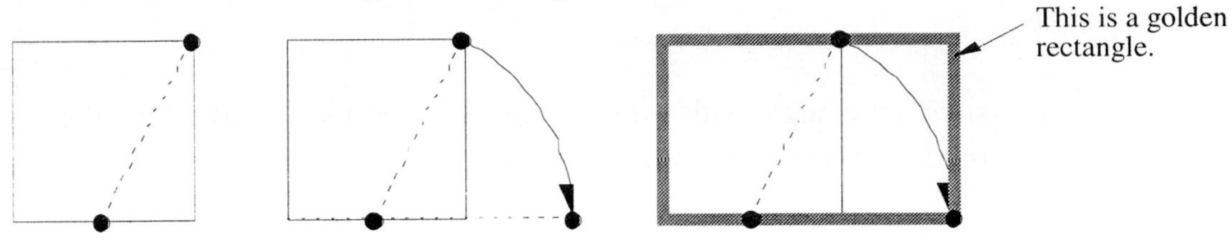
This is a golden rectangle.

Draw your golden rectangle here:

What is the slope of the diagonal of your golden rectangle?

TASK 6: Using the golden rectangle to analyze a design.

On a piece of tracing paper draw a golden rectangle and its regulating line. As you did in Task 3, use these regulating lines to find golden rectangles in the design of S. Giorgio Maggiore (drawing attached). How many can you find? Be creative. You'll find them throughout the design.

VI. WHAT IS SO SPECIAL ABOUT THE NUMBER 1.618033989....?

TASK 7: Investigate.

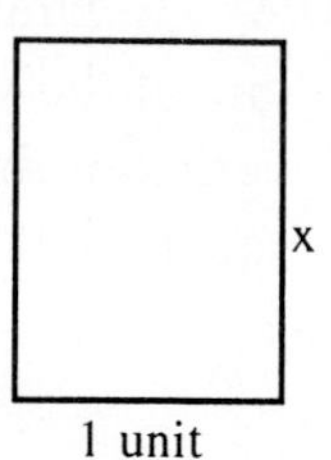

If we try to find a sequence of rectangles which are both geometric and additive, we need to consider the "seed" rectangle and then build the sequences.

Consider a rectangle with width 1 unit. We don't know how long the rectangle should be so for now we'll call it *x*.

Geometric sequence: 1, x, x^2, x^3, x^4, x^5, x^6, ... (multiple is ________)

Additive sequence: 1, x, x+1, 2x+1, 3x+2, 5x+3, ...

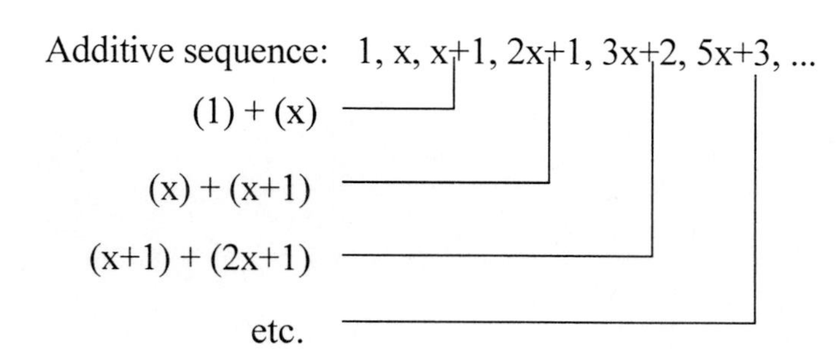

If we want these two sequences to be the same, then x^2 must be equal to x+1. (In fact, x^3 must be equal to 2x+1, x^4 must be equal to 3x+2, and so on). Set up this equation and solve it:

$$x^2 = x+1$$

The solution of this equation will give us the dimension required to create the "seed" rectangle. Notice that you got two answers. We'll only use one of them. Which one should we use and why?

(Alternately, you could have set up any of the required equalities and solved those equations. Try using your graphing calculator to solve $x^3 = 2x+1$ and $x^4 = 3x+2$.)

TASK 8: Conclusion.

Create the first 10 terms of both the geometric sequence and the additive beginning with 1 and 1.618033989...

	Geometric Sequence	Additive Sequence
1)	____________	____________
2)	____________	____________
3)	____________	____________
4)	____________	____________
5)	____________	____________
6)	____________	____________
7)	____________	____________
8)	____________	____________
9)	____________	____________
10)	____________	____________

This is the only sequence of positive numbers that is both additive and geometric.

The positive solution of the equation $x^2 = x+1$ (ie. 1.618033989...) is considered so important in mathematics, art and architecture that it has been given the symbol φ (the Greek letter phi).

Sketch the "seed" rectangle. Measure it as accurately as you can. What is the slope of the diagonal of any of the rectangles generated by this sequence?

VII. EXTENSIONS OR ADDITIONAL ACTIVITIES

1) Investigate other sequences of rectangles.
   a) Create a sequence of rectangles based on *arithmetic sequences.* An arithmetic sequence is a list of numbers created by adding the same number to each term to create the next one. E.g. 1, 4, 7, 10, 13, 16, ...

      What is the limit of the sequence of the slopes of the diagonals?

   b) The *dynamic rectangles* are a sequence of rectangles created by drawing the diagonal of a square and using it for the base of the second rectangle. Use the length of the diagonal of the second rectangle as the base for the third rectangle and so on.

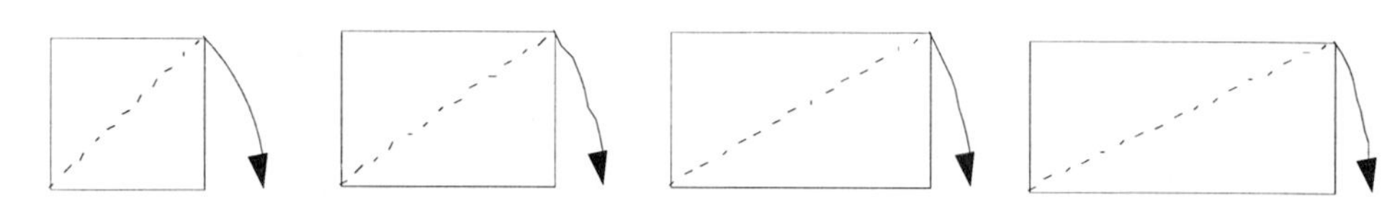

      Calculate the lengths of the bases of this sequence of rectangles using the Pythagorean Theorem. Find the limit of the slopes of the diagonals of the dynamic rectangles.

2) Construct a golden rectangle using the procedure in Task 6. Prove that it really has the correct proportion without measuring it. (Hint: Start with a 1" by 1" square and calculate the lengths of the rest of the line segments used.)

3) Research the architect LeCorbusier to find out how important proportioning systems were in his work.

4) Research the Fibonacci numbers. What are they and how do they fit into this investigation?

5) Choose a few consecutive rectangles from an additive sequence. Make some copies of each of them and use them as tiles to create geometric patterns. What do you notice about how they fit together?

Ca D'Oro: Venice , Giovanni & Bartolomeo Buon. (Drawing by Jeffrey Paquin.)

S. Giorgio Maggiore, Venice, Andrea Palladio. (Drawing by Jeffrey Paquin.)

MATHEMATICS LABORATORY INVESTIGATION

GETTING A CHARGE OUT OF MATH

Topics: **EXPONENTIAL FUNCTIONS, SHIFTING GRAPHS, NONLINEAR REGRESSION**
Prerequisites: *Linear Regression*
Equipment: Analog voltmeter, digital voltmeter, 27,000 μf capacitor, 2 long/2 short banana to banana leads, 2 banana to alligator leads, 220 ohm resistor, a terminal block, 6V battery, stopwatch or second-sweep hand

I. INTRODUCTION

During a weather forecast a meteorologist will often refer to the barometric pressure reading. This reading, often stated in inches of mercury, has little meaning by itself. During the forecast the weatherman will add whether the barometer gauge needle is rising or falling. What's important in order to predict fair or worsening weather is to determine the *trend* of the barometric pressure, not just an isolated reading. A barometer with a needle which can indicate any reading along the continuum of the dial face is an example of an *analog device*.

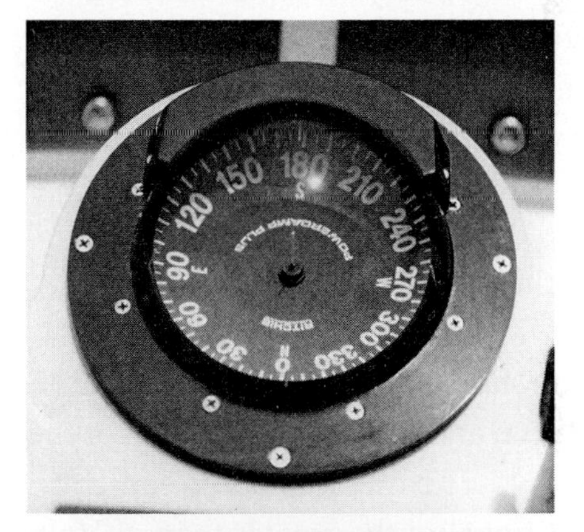

There is a common misconception for one to look at a dial analog instrument and dismiss it as old, outdated and inaccurate. Conversely, there is a tendency to view instruments with digital LCD (liquid crystal diode) displays as modern, accurate and the best style of instrument to use. In practice each type of meter has its advantages and limitations.

An analog verses digital analogy can be made between video and photographs; video (analog) and photographs (digital) continue to co-exist. A video camera provides us with a continuous flow of vision; while the camera captures one discrete frame at a time. The video shows us what happens next or what happened before - we have no way of knowing this by looking at a photograph. *Analog data is useful for determining trends* - and *predicting* what might happen next. This advantage doesn't eliminate the need for pictures. A photograph is still useful. To analyze a detail in the scene, the photograph freezes the moment and allows a view that otherwise might pass by too quickly to be observed. To gather the data from a specific instant in time a digital meter is very useful.

II. THE EXPERIMENT

To illustrate the advantages and disadvantages of analog and digital instruments, something must be measured. In this lab the voltage across an *electrical storage device* known as a *capacitor* will be measured. A capacitor in an electrical circuit is like a water tank: it can be filled, hold its contents, and be emptied.

A multimeter will be used to measure the voltage during the process of charging (or "filling") the capacitor . Then the multimeter will measure the rate of discharging (or "emptying") of the same capacitor. A graph of the data will illustrate the relationship of voltage over time.

Task 1: Preparing the equipment.

The initial circuit is composed of a battery or power supply (V), and a capacitor (C). The two battery terminals should have clip wires (S) attached which can be switched on and off from the capacitor terminals.

Prior to building this circuit, test the function of the *digital* and *analog* voltmeter and circuit components. Clip the meter to the positive and negative terminals of the battery. The meter should read close to, if not slightly greater than, 6 Volts.

Record the actual maximum voltage. ____________

Clip the meter to the positive and negative terminals of the capacitor. It should read zero. If not, fully discharge the capacitor by briefly connecting positive and negative capacitor terminals. Test again with the meter. When you are confident the capacitor is fully discharged, proceed.

Task 2: Charging and discharging the capacitor.

Once again attach a voltmeter to the capacitor. The voltage reading on the meter should be 0. Refer to the diagram 1 at the end of the lab as you set up the initial circuit. Connect a black lead from the negative terminal of the battery to the negative terminal of your capacitor and a red lead from the positive terminal of the battery to the positive terminal of the capacitor. As the capacitor is being charged, watch the values change until the meter reads approximately the maximum value recorded earlier.

Disengage the battery by removing the positive lead from the circuit - the meter will still read the maximum value recorded before (the stable voltage of approximately 6 volts across the capacitor).

Keep the meter attached to the terminals of the capacitor while you discharge it. **To avoid damaging the capacitor during its discharge cycle, "short out" the capacitor by disconnecting the battery ends of the alligator leads and by connecting the leads from the capacitor terminals together.** Observe the decreasing voltage values which approach zero once again.

This cycle can be performed over and over by alternately using the battery to charge the capacitor, then shorting the capacitor. Watch the needle of the meter as the capacitor is charged and discharged. Did the capacitor charge quickly or slowly? ___________
Did the capacitor discharge quickly or slowly? ___________

III. ADAPTING THE EXPERIMENT TO EXPLORE THE NATURE OF THE RATE OF CHARGE AND DISCHARGE

By adding resistance to the initial circuit, the time it takes to either charge or discharge the capacitor can be slowed down and measured.

Here's a schematic drawing of the circuit.

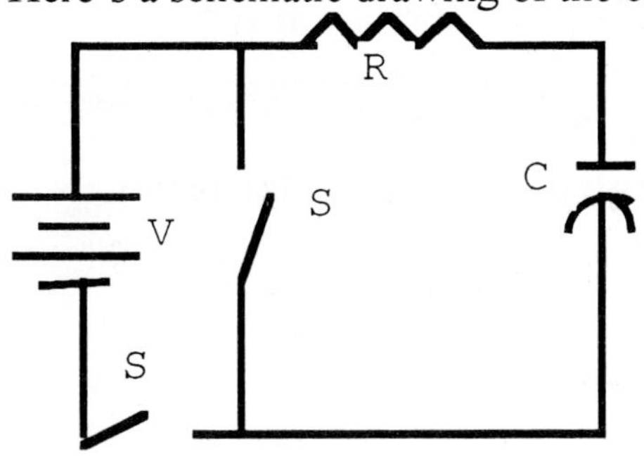

Task 3: Add resistance to the initial circuit by attaching the resistor between the battery and the capacitor. (Refer to diagram 2 at the end of the lab).

Task 4: Recording the observations from the analog meter.

Using the new circuit, focus your attention on the analog meter. For each of three separate trials, **time and record below** how long it takes for the capacitor to be fully charged and then for the capacitor to fully discharge. Be sure to short the circuit over the resistor , not the battery by disconnecting the lead from the positive terminal of the battery and connecting that lead to the negative terminal of the capacitor. (Refer to diagram 3 at the end of the lab.) Then **record the average** of the three trials.

	Total Cycle		
	Trial 1	Trial 2	Trial 3
Charging Cycle			
Discharging Cycle			

Do these calculations.

	Average Time
Charging Cycle	
Discharging Cycle	

The *charge/discharge rate* is the change in the voltage over time.

How does the needle on the meter behave near either end of a cycle?

During the charging cycle does the <u>voltage</u> increase or decrease?

Is the <u>rate</u> constant or varying?

Will a graph of the data be linear or non-linear?

Discuss the nature of the discharging cycle in similar terms.

Task 5: Predicting the graphs of voltage over time using the analog meter.

From your observations, **sketch** your prediction of the graph of voltage over time for the charging cycle.

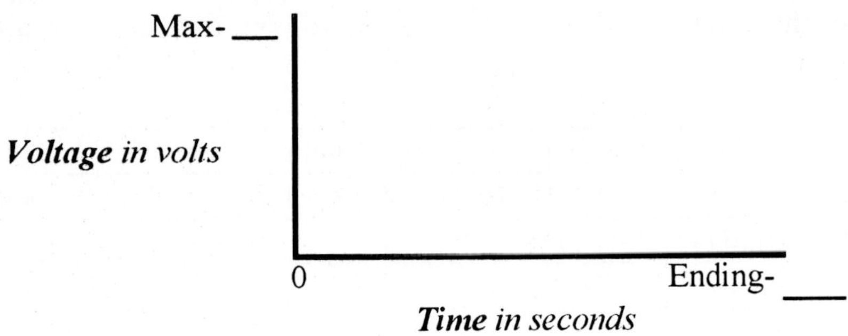

Now **sketch** the discharge cycle.

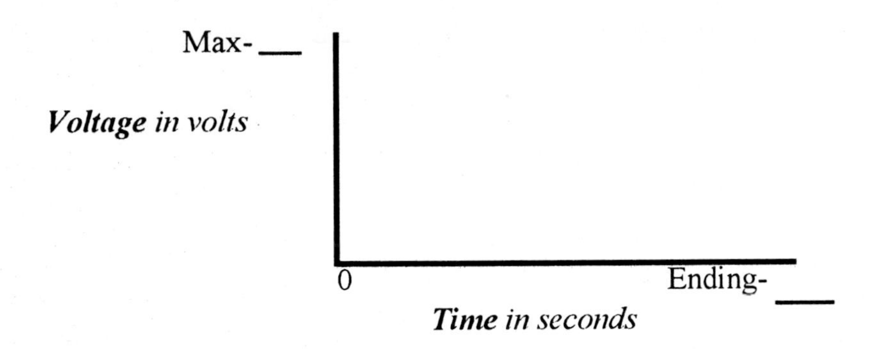

Task 6: Predicting the graphs of voltage over time using the digital meter.

Next, **focus on reading the digital meter as you charge and discharge the slowed down circuit**. Do these readings give any indication of the nature of the rate? They may or may not, but the digital meter will make it easier to collect "snapshot" readings at specific times.

The frequency of the readings needs to be determined. Since the prediction before was a "slightly informed" guess, it is still hard to tell how many data points are necessary to make a good prediction.

Some extra thought is needed in deciding when the capacitor is fully charged as you read the voltage from the digital meter. Since the size of the capacitor has not changed, you can use the average charging time recorded from the analog meter as a guideline in reading the digital meter.

Once again time and record the length of the charging cycle. ______

What guidelines should be set for identifying the end of the cycle?

Perform three trials for each cycle and record your data below. Calculate **the average time** for each cycle and record it below.

	Total Cycle		
	Trial 1	Trial 2	Trial 3
Charging Cycle			
Discharging Cycle			

Do these calculations.

	Average Time
Charging Cycle	
Discharge Cycle	

Consider your average charging time. **What is half of that time?** ________
Record this number under elapsed time in the middle row of the chart below.

Once again charge and discharge the capacitor taking three readings of the voltage for each of the cycles - one at the beginning of the cycle, one halfway **(in time)** through the cycle, and one at the end of the cycle.

Charging Cycle

	Elapsed Time	Voltage
Initial		
Halfway		
Final		

Plot these three points. Draw the curve.

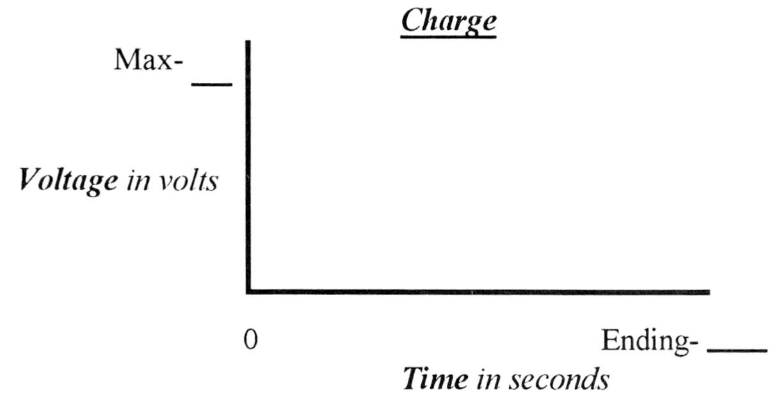

Discharging Cycle

	Elapsed Time	Voltage
Initial		
Halfway		
Final		

Plot the point. Draw the curve.

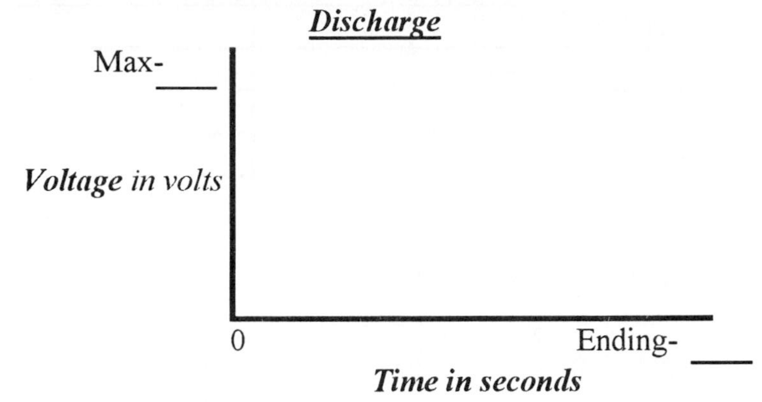

Are these plots consistent with your prediction using the analog data?

Task 7: Extending the data.

Three points do not give much confidence in predicting further data. **Repeat the experiment**, but **increase the number of *interim* datapoints** to a minimum of five. **Use equal time intervals** and record them in the chart below. Then **read and record the voltage** at each point in time.

Charging Cycle

	Elapsed Time	Voltage
Initial		
Halfway		
Final		

Discharging Cycle

	Elapsed Time	Voltage
Initial		
Halfway		
Final		

Plot the data for the charge and the discharge cycles. Include a scale and the units on each axes.

IV. ARRIVING AT CONCLUSIONS

Task 8: Plotting the points for the Discharge Curve on a graphing calculator.

Use your graphing calculator to plot the two sets of points for the Charge Curve and Discharge Curve. Use a different symbol to indicate the points on each curve.

Compare your plots with each member in your group.

Task 9: Finding an appropriate regression equation for the Discharge Curve with a graphing calculator.

Use only the points from Discharge Cycle and enter the equations for a variety of regressions into your ***Calculator*****.** As you write the different regression equations, use V for the voltage and t for the elapsed time. Approximate all constants to the nearest thousandth.

Write the equation for the linear regression.

____

Write the equation for the quadratic regression.

____

Write the equation for the cubic regression.

____

Write the regression for the exponential regression.

____

Which regression equation seems to have the best fit?

____

Did other members of your group get the same equation?

If not, how do you explain the differences?

V. EXPLORING THE EXPONENTIAL FUNCTION

A basic *exponential function* has the form $\mathbf{f(x) = b^x}$ where b is a constant greater than zero and not equal to one. That allows many choices for the base. In engineering applications as well as in growth/decay applications, the base most often chosen for use is *e* (e = 2.7181...), after Leonhard Euler, one of history's greatest mathematicians. It is such an important number that there is a special key on your calculator, called e^x. When providing an equation for the exponential regression, graphing calculators make use of many different bases but do not use base e. There are conversion formulas that will convert exponential functions written using a base of e into an exponential function using another base or vice versa.

The general shape of the exponential function is not dependent on its base. Since this lab deals with engineering applications of exponential functions, the next several tasks will have you explore the variations of the decreasing function $y = e^{-x}$. (Note that since most applications are based on elapsed time, the independent variable being used is t, not x.)

Task 10: Graph $y = e^{-t}$, $y = e^{-2t}$, $y = e^{-4t}$ on your grapher.

What do these curves have in common?

Does the shape of $y = e^{-t}$ from $t \geq 0$ **resemble** either the graph of charge or the graph of your discharge curve?

A *mirror image* through the vertical axis, called a *reflection* of the graph, can be developed by negating the exponent of the function. The *reflection* of curve through the horizontal axis, also called a *mirror image*, can be produced by negating the entire right side of the function.

Task 11: **Graph** $y = e^{-t}$, and its reflections, $y = e^{-(-t)}$ and $y = -(e^{-t})$ on your grapher.

As you **reproduce the graphs** below, **write the equation** for each curve by its graph.

Reproduce the reflection over the x-axis here. **Reproduce** the reflection over the y-axis here.

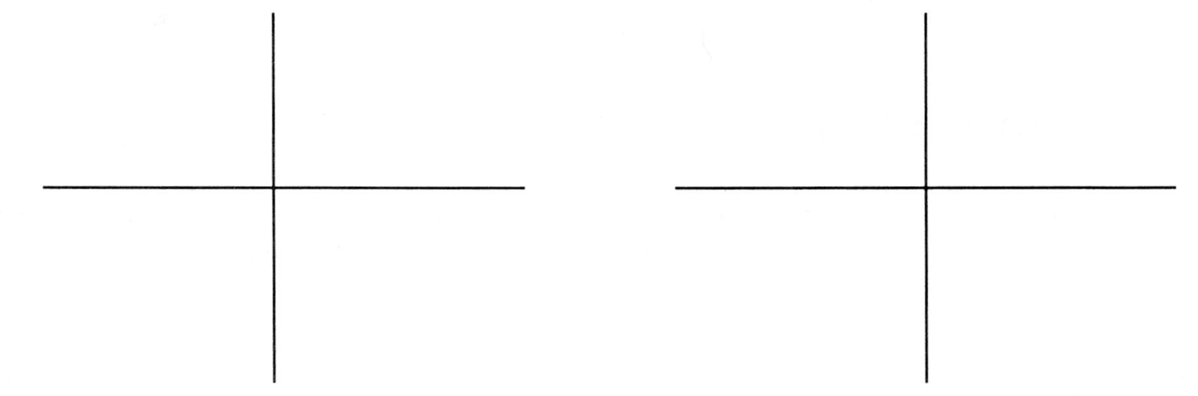

Which reflection has the same shape as your prediction for Charge Curve?

As you compared the shapes, did you notice that the position of the curves on the axes were different? These changes in position, or shifts, can be produced by adding (or subtracting) values in the function.

Task 12: Shifting the graph of $y = e^{-t}$.

Graph $y_1 = e^{-t}$, $y_2 = e^{-(t - 2)}$ and $y_3 = e^{-(t + 1)}$ on your grapher.

Describe how y_1 has been shifted to produce the graph of y_2.

Describe how y_1 has been shifted to produce the graph of y_3.

Do these changes appear to cause vertical or horizontal shifting of the curves? ________________

Graph $y_1 = e^{-t}$, $y_2 = e^{-t} + 1$, $y_3 = e^{-t} - 2$.

Describe how y_1 has been shifted to produce the graph of y_2.

Describe how y_1 has been shifted to produce the graph of y_3.

Do these changes appear to cause vertical or horizontal shifting of the curves? ________________

Describe how to adjust the input or output variable of a function to create a horizontal and/or a vertical shifting of a function.

The Charge Curve can be written as a reflection and a shifting of an exponential function. Look back to the your regression equation for the Discharge Curve. It was written without using the shifting adjustment but by using a stretching (shrinking) adjustment. Thus the equation is written in the form $y = a * b^x$.

Rewrite this function using a base greater than one and a negative exponent.

____

Task 13: Deciding the limiting value.

Once again look at the graph of $y = e^{-t}$.

What value does y have when $t = 10$? ________

What value does y have when $t = 25$? ________

What value does y approach as t gets larger and larger? ________

This is called the *limiting value* for the curve.

Now shift $y = e^{-t}$ up 2.5 units. **What is the limiting value** of the new curve?

____

In practice, electrical engineers will determine the limiting value indirectly. The limiting value of an experiment is the height of the horizontal asymptote of an exponential function.

For the charging cycle of either circuit, **what was the limiting value**?

If voltage is a function of time, what is the **equation of the horizontal asymptote** for the Charging Cycle? ________________________

VI. APPLICATIONS OF $y = 1 - e^{-x}$

There are many applications in engineering and growth/decay situations that make use of a variation of the exponential function. It is of such importance that it really should have its own name. Let's call it the *Constrained Growth Curve*. One such application occurs in the curing of concrete.

Task 14: Making predictions using the Concrete Curing Equation.

Consider the job of curing concrete. (Curing concrete refers to the completion of a chemical reaction.) The equation $P = 100(1 - e^{-0.16x})$ will provide a reasonable estimate of the percentage of concrete cured after x days.

What percentage of concrete will be cured after the first week? ________

What percentage of concrete will be cured after the third week ? ________

In how many days will the curing be 50% complete? ________

VII. HALF LIFE AND TIME CONSTANTS

Task 15: Refer back to the Discharge Curve.

The *half life* for the Discharge Curve is the time it takes the voltage to reach half the value of its initial voltage. **Calculate** the half life. ___________

In practice electrical engineers are more interested in the time constant rather than the half life of a Discharge Function. This time constant can be calculated directly from the values of the circuit components. It occurs as $y = e^{-t}$ approaches its limiting value, when the voltage is 37% of its original value ($e^{-1} = 0.37$).

Calculate the time constant for your discharge data. ___________

How does this compare to the **theoretical value,** which is **(R*C)** where R is measured in ohms and C is measured in farads? (Refer back to the size of the capacitor and the resistor in the equipment list on page one.)

VII. SUMMING UP

While plotting the Charging/Discharging Cycles, you have been graphing exponential curves. In latter sections of this lab you have seen some equations that describe exponential functions and their corresponding graphs. **Describe the distinguishing features of an exponential function.**

In an electronics lab, you will commonly find both an analog and a digital meter. **Why?**

Which of the distinguishing features of an exponential function was illustrated using an analog meter?

Which of the distinguishing features of an exponential function was illustrated using a digital meter?

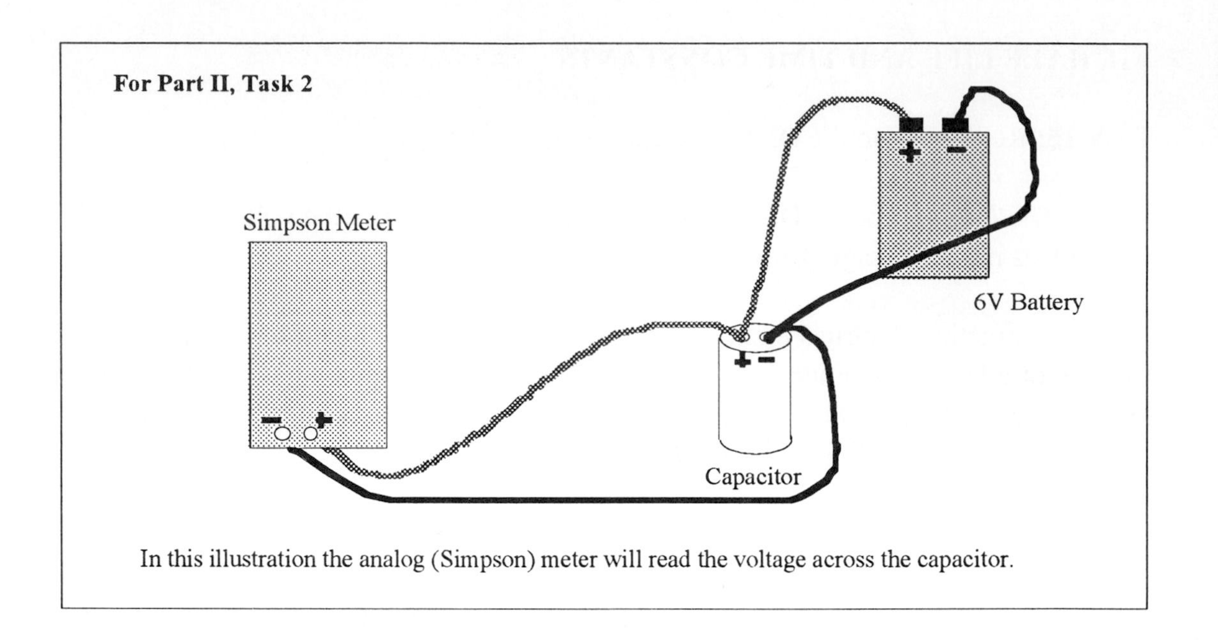
For Part II, Task 2
Simpson Meter
6V Battery
Capacitor
In this illustration the analog (Simpson) meter will read the voltage across the capacitor.

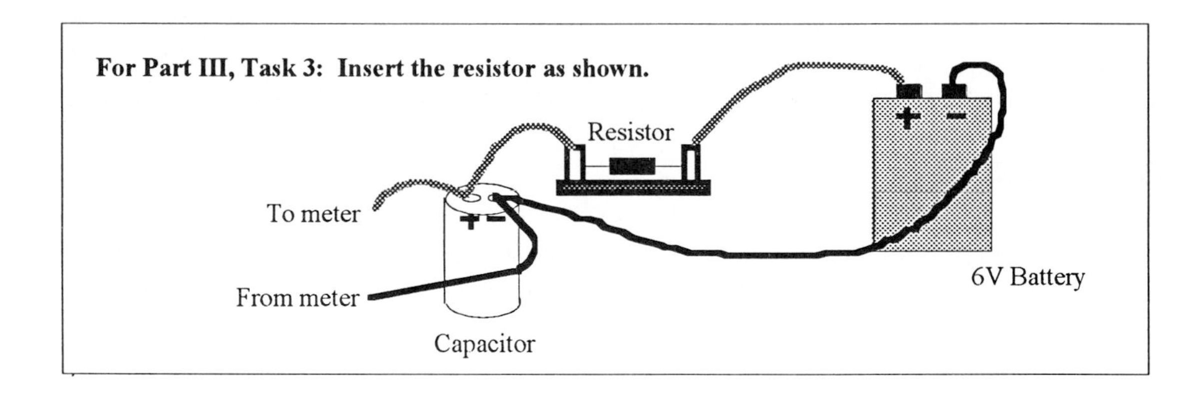
For Part III, Task 3: Insert the resistor as shown.
Resistor
To meter
From meter
6V Battery
Capacitor

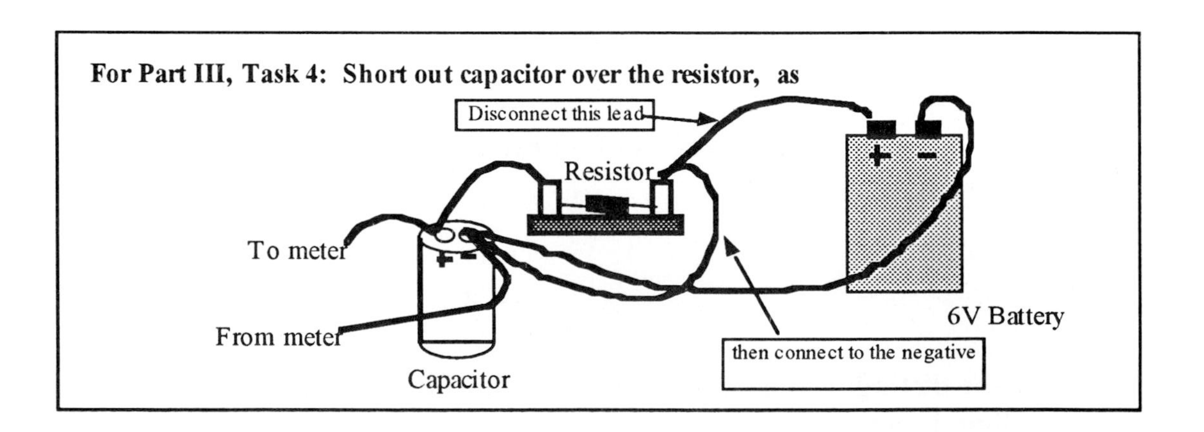
For Part III, Task 4: Short out capacitor over the resistor, as
Disconnect this lead
Resistor
To meter
From meter
6V Battery
then connect to the negative
Capacitor

MATHEMATICS LABORATORY INVESTIGATION

LEASE SPACE ANALYSIS

Topics: **AREAS, RATIOS**
Prerequisites: *Geometry, Pythagorean Theorem*

I. INTRODUCTION:

When leasing space, a company must analyze many factors in order to be able to compare spaces in different buildings. In this lease space analysis, we will be looking at issues relating to area only. In a typical lease space analysis, you would consider many other (intangible) variables as well (such as the location of the building, views, neighborhoood stores, or rent). We will be focussing only on comparisons that relate to size, floor plate (the building's basic shape) and efficiency.

Size
The tenant will pay the landlord based upon the rentable square footage (area). This is based upon the overall gross square footage, minus shafts and vertical penetrations, plus a factor added in by the landlord to cover common areas, such as lobbies and corridors. Thus, the rentable area is generally a constant. It cannot be physically measured, negotiated or changed in the leasing process.

Floor Plate
Another important consideration in determining a building's suitability to house a particular business is the structure of the firm desiring to use the space. Organizational structure and heirarchy have a direct relationship with an organization's function and its need for a specific type of space. Insurance companies, for example, tend to be open-office intensive, and would require a large floor plate with large depth from windows to core space. A law firm, on the other hand, tends to be private-office intensive and would make more efficient use of a smaller floor plate and leasing depth.

Efficiency
When leasing space, one must first determine the correct amount of space (not too much, not too little). Usually, buildings come in various shapes and sizes, which are often difficult to compare just by looking at the

shape or the number of rentable square feet. The efficiency, a measure of how much of the space is usable, is a major consideration. Some shapes are inherently more efficient than others. In other buildings, mechanical rooms, utility or corridor spaces will shrink the usable square footage. It is very important to distinguish the usable area from the rentable area in order to make building comparisons.

II. SITUATION

A corporation is interested in moving into the Boston area. There are two properties it is considering. Among the considerations (location, parking, views, lobbies, convenience, etc.) there will be some quantitative information describing the properties. One such piece of information is the area of the space to be leased. The two properties it is considering are 222 Berkeley Street and 2 International Place. In the following pages are two floor plans of the each of the two properties. The larger floor plan of each property is a 1/16" scale, that is 1"=16'.

III. APPROXIMATING THE AREAS

TASK1: Estimating the area of 222 Berkeley Street

Using your 1/16" scale drawing, approximate the area of 222 Berkeley Street (including the elevator shafts, bathrooms, etc).

Explain how you arrived at your estimate.

Estimate the area of the central region containing the elevator shafts, staircases, mechanical closets, and bathrooms.

Discuss the accuracy of these computations.

TASK 2: Preliminary estimate of the area of 2 International Place

Whereas 222 Berkeley Street contained right angles and was composed of rectangles, 2 International Place is more "circular". Given this, our first estimate of the area will be made by approximating the floor plan with a circle.

If you were to approximate the floor plan by a circle, what value would you use for the radius?

Using that circle, estimate the area.

Explain whether you believe your estimate to be high or low.

TASK 3: Refining the area of 2 International Place

Rather than using a circle to approximate the area, we will use a regular 28 sided polygon. To do this, we need to know the formula for the area of a regular polygon. As an example, consider a regular pentagon (5 sides). Draw one.

Assuming that r denotes the distance from the center of the pentagon to a vertex and that s denotes the length of a side, write a formula for the area.

Hint: Draw a triangle using the central point and two adjacent vertices. (Its sides will have length r, r, and s.) Call the altitude of the triangle a, and find it first.

Do the same for the regular octagon.

Using the same method, find a formula for the area of the regular 28 sided polygon.

Use the formula from above to estimate the area of 2 International Place. Notice that the shape is not precisely a regular polygon. There is one place (at the top of your plan) where the building has a bit of an indentation. Assume, for the moment that the indentation has been "filled in" by four more sides. On the bottom of the plan, there are two small triangular regions which jut out from the polygon. Again, remove these triangular regions and assume that they have been replaced by a straight edge. This will give 28 sides.

Make a second estimate, correcting the first, by considering these irregularities.

Compare your answers with those of other members of your class. What differences do you see? Based upon those comparisons, do you think your estimates are high, low, or neither?

TASK 4: USABLE SQUARE FOOTAGE

The area estimates you have calculated above are the Rentable Square Footage (RSF). Not all of this space is usable. As you can see from the floor plans, there are maintenance and utility areas (including elevators, stairs, HVAC, etc.) in the center of each floor. The Usable Square Footage (USF) is defined to be the RSF minus the area of these unusable regions.

Using your computation from Task 1, calculate the USF for 222 Berkeley St.

Calculate the USF for 2 International Place.

TASK 5: DETERMINING THE EFFICIENCY

The ratio of USF to RSF is called the *efficiency*. Calculate the efficiency of both properties and answer the following questions.

1. Which property is more efficient?

2. What does this mean?

3. What percent of space in each property is unusable? (This is called the *Loss Factor*.)

4. What is the possible range of values for the efficiency?

5. Consider the less efficient property. What fraction of its unusable space would have to be converted to usable space to make it as efficient as the more efficient property?

TASK 6: DETERMINING THE RENTAL COSTS

The rent for 222 Berkeley St. is $35 per square foot of RSF. The rent for 2 International Place is $40 per square foot of RSF. In order to compare the rents of the two properties, the rents must be recomputed in terms of dollars per square foot of USF.

Find the rent for 222 Berkeley Street.

Find the rent for 2 International Place.

Recompute the rents in terms of dollars per square foot of USF.

TASK 7: DETERMINING THE PERIMETER TO AREA RATIO

The Perimeter to Area Ratio of a space indicates how tightly clustered the usable space is. For example, 1600 square feet of space could take the form of a 40' by 40' square or of a 20' by 80' rectangle. In the former, no two points in the space are more than 56' (the diagonal of the square) apart, while, in the latter, distances of greater than 80' are clearly possible.

The Perimeter to Area Ratio is expressed in the form

1 linear foot: n square feet

The larger n is, the more densely packed the space is.

For each property, compute the perimeter of the floor plan. Then, compute the Perimeter to Area Ratio by dividing the perimeter by the RSF. Which space is more tightly clustered?

Generalize about the effect of the Perimeter to Area Ratio. (How does it affect window accessibility, for example?)

Extensions:

1. Find the formula for the area of a regular polygon with n sides.
2. Show that, as n gets large, the area of this polygon approaches the area of a circle.
3. Assume that a single floor of 222 Berkeley Street is to be occupied by several different departments of the leasing firm. Each department's budget is to be charged for its portion of the rental costs. Choose a section of the floor, compute its usuable area and then, using the efficiency computed in Task 7, compute the rental area for which the department should be charged. (Rental Area = Usable Area / Efficiency)

GLOSSARY:

Gross Area of Floor

This is determined by measuring all the space on the floor to the inside finished surface of the exterior walls (or windows).

Loss Factor

The difference between the rentable and usable square footage, expressed as a percentage of rentable area. This is used to indicate the amount of space devoted to common areas of the building, and includes their core and shell dimensions and thicknesses.

Perimeter to Area Ratio

Measured as 1 linear foot: n square feet, this is a measure of the density of the building. A high value for n implies that a larger area is enclosed by a smaller perimeter, which indicates a higher density of the space.

Scale

The ratio of the size of a drawing (usually in inches) to the actual size of the builing being drawn (usually in feet).

Usable Square Footage (single tenant floor)

Subtract from the Gross Area of the floor the following:

- Public elevator shafts and elevator machines including their enclosing walls
- Public stairs and their enclosing walls
- Heating, ventilating and air conditioning facilities (including pipes, ducts, and shafts) and their enclosing walls.
- Main telephone equipment rooms and main electric switch gear rooms.

MATHEMATICS LABORATORY INVESTIGATION

LIFE CYCLE COST ANALYSIS

Topic: *Introduction to Exponential Functions, Present Value*

Prerequisite: *Graphing Calculator, Linear Functions*

I. INTRODUCTION:

The evaluation of cost as it relates to the purchase of capital equipment is a complex one. It involves more than the simple purchase and delivery price of a particular item. Even when that item is one which is routinely stocked by a supplier, the actual cost of using that item, instead of a comparable one from a different supplier, is a function of many factors. "Life-Cycle Cost Analysis" is a way to evaluate as many of the cost functions as possible so that decision makers can base a purchase decision on the maximum amount of available, relevant data.

If, for example, a company were interested in providing an automobile for a salesperson to use, the company would have several choices to make. It could lease a vehicle for a fixed period of time or it could buy a vehicle outright. If it were to purchase a vehicle, it could buy an expensive vehicle, which might come with a long-term maintenance agreement as part of the purchase price, or it could buy a less expensive vehicle and pay for whatever maintenance was required on an as-needed basis. A third option might be to require the salesperson to buy the vehicle and pay the salesperson a fixed rate per mile for business use of the vehicle.

The situation is similar for long term facility investments. For example, when specifying an upgraded air conditioning system for a high rise building, it would be necessary to consider factors such as:

- initial purchase price
- installation costs
- expected life of the unit
- the company's cost of capital (ie. what interest would the company have to pay to borrow the money for the AC unit)
- replacement costs for parts
- labor costs of repairing the unit
- on-going or scheduled maintenance of the unit
- the value that the AC unit adds to the building (if owned by the company)

In environmental engineering, the issue could involve the selection of pumps at a wastewater treatment plant. One option could include, for example, the purchase of a single, large capacity, variable speed pump that can operate over a very large range of flows. An alternative plan could involve the purchase of three smaller, fixed speed pumps so designed that one pump could handle the low flows expected and all three together could handle the peak flows expected. By either varying the speed of the one variable speed pump or by turning fixed speed pumps on or off, the plant operator could manage to deal with all flows effectively under either scenario.

The cost issues are the higher maintenance costs of the variable speed pump versus the high reliability (with consequent lower maintenance costs) of the fixed speed pumps and the lower initial cost the single pump versus a higher overall initial cost for the three fixed speed pumps together. Additional considerations could include the longer life of the fixed speed pumps, thereby requiring less frequent replacement, and fewer employee hours being required to perform routine maintenance tasks on the fixed speed pumps even though there are more of them to maintain.

As an additional example, the City of Boston had to make a decision regarding the Long Island Bridge. The bridge was in need both of structural repairs and of deleading and repainting. The question faced by the City was: should all the painting and repairing be done at once or would it be more cost effective to do only the repairs (which had to be done for safety reasons) and put off the painting for an additional year? The total cost would be higher to spread the job over two years as certain costs would need to be duplicated (eg. mobilization of crews and material, staging for access under the bridge, containment for either deleading or structural repairs, inspections). There would be financial advantages, however, to deferring some of the cost until the next fiscal year.

In examples such as these, each of the options has a different initial cost, a different annual maintenance cost, and a different salvage value, or trade-in value, at the end of any specific period of time. Moreover, the length of time that each option would reliably provide a solution to the problem would differ. The evaluation of those options on a "present value" basis is the crux of life-cycle costing.

II. SITUATION:

A small firm is looking to buy some furniture for its lobby. Its design team has identified two acceptable brands, Chairs R Us and Elegance Unlimited. Chairs R Us has the advantage of price, their furniture costing $10,000 as opposed to $14,000 at Elegance Unlimited. Both companies will deliver and install the furniture for the same price, $1,000. The furniture from Elegance Unlimited has an expected lifespan of eleven years, whereas the furniture from Chairs R Us has a lifespan of only eight years. Furthermore, the Elegance Unlimited furniture requires less maintenance ($1200/year versus $1500/year from Chairs R Us). Our goal is to make a determination as to the cost effectiveness of these two options.

TASK 1: COMPARING THE TOTAL COSTS

The simplest method of comparison is to compare the total costs of the two options. This can be easily viewed using charts. For example, the total cost for the Elegance Unlimited furniture is exhibited in Figure 1.

Elegance Unlimited Furniture												
	Initial Costs											
Purchase Price	$14,000											
Installation/Delivery	$1,000											
Total Initial Cost	$15,000											
Annual Maintenance		Year 1	Year 2	Year 3	Year 4	Year 5	Year 6	Year 7	Year 8	Year 9	Year 10	Year 11
Labor		$900	$900	$900	$900	$900	$900	$900	$900	$900	$900	$900
Material		$300	$300	$300	$300	$300	$300	$300	$300	$300	$300	$300
Annual Total Maint.		$1,200	$1,200	$1,200	$1,200	$1,200	$1,200	$1,200	$1,200	$1,200	$1,200	$1,200
Total Annual Costs	$13,200											
Total Cost	$28,200											

Figure 1

In this chart, we have shown the initial costs as well as the continuing costs over the expected eleven year lifespan of the furniture. The continuing maintenance costs have been broken down as $900 in labor and $300 in materials. The total cost over the furniture's lifespan is shown at the bottom.

Create a similar chart for the Chairs R Us furniture by filling in figure 2 below. The lifespan will need to be decreased to 8 years and the various costs adjusted. You may assume that the annual costs break down as $1100 for labor and $400 for materials.

Chairs 'R Us Furniture												
	Initial Costs											
Purchase Price												
Installation/Delivery												
Total Initial Cost												
Annual Maintenance		Year 1	Year 2	Year 3	Year 4	Year 5	Year 6	Year 7	Year 8			
Labor												
Material												
Annual Total Maint.												
Total Annual Costs												
Total Cost												

Figure 2

By comparing the information contained in these spreadsheets, explain which option appears to be the most cost effective and why?

TASK 2: PRESENT VALUE METHOD

One obvious disadvantage of the previous method is that you are comparing expenses being made in the future. If you know that you are going to have an expense of, say, $1200 eight years from now, you could put a smaller amount of money aside now (in an interest bearing bank account, for instance) and let it accumulate for eight years. At that point, there should be enough money available to pay the expense. Viewed from this perspective, a $1200 cost eight years in the future is really a smaller cost if paid today. This smaller cost is the basis of the concept of *Present Value*.

The uncertainty in the present value computation arises from what the "interest rate" should be. We clearly do not know what will happen to the economy in the future, so we need to make an educated guess. Such an approximation is available from financial institutions (and from anyone else who cares to venture such a guess) and is called the *Discount Rate.*

For example, if we assume a discount rate (interest rate) of 7%, an investment of approximately $698 would yield $1200 in eight years. We say that the present value of the $1200 (in 8 years at a 7% annualized percentage rate (APR)) is $698. In general, for an amount of money, A, to be spent t years from now, we can compute the present value (PV) as

$$PV = \frac{A}{(1 + DR)^t} = A(1 + DR)^{-t}$$

where DR is the discount rate (.07 in this example).

Figure 3 modifies the chart given in the Total Cost Comparison by recomputing the future costs in terms of present value (Equivalent Annual Cost). The sum of the initial costs and the present values gives the *Net Present Value*.

Elegance Unlimited Furniture												
Discount Rate	7%											
	Initial Costs											
Purchase Price	$14,000											
Installation/Delivery	$1,000											
Total Initial Cost	$15,000											
Annual Maintenance		Year 1	Year 2	Year 3	Year 4	Year 5	Year 6	Year 7	Year 8	Year 9	Year 10	Year 11
Labor		$900	$900	$900	$900	$900	$900	$900	$900	$900	$900	$900
Material		$300	$300	$300	$300	$300	$300	$300	$300	$300	$300	$300
Annual Total Maint.		$1,200	$1,200	$1,200	$1,200	$1,200	$1,200	$1,200	$1,200	$1,200	$1,200	$1,200
Equivalent Annual Cost		$1,121.50	$1,048.13	$979.56	$915.47	$855.58	$799.61	$747.30	$698.41	$652.72	$610.02	$570.11
Total Equiv. Annual Costs	$8,998.41											
Net Present Value	$23,998.41											

Figure 3

For the Elegance Unlimited furniture (as shown in figure 3), write the formula for the Present Value of the annual maintenance costs as a function of time. Use this to verify several of the entries in the "Equivalent Annual Cost" row of figure 3.

Copy the data from figure 2 into the chart below.

Chairs 'R Us Furniture												
Discount Rate	7%											
	Initial Costs											
Purchase Price												
Installation/Delivery												
Total Initial Cost												
Annual Maintenance		Year 1	Year 2	Year 3	Year 4	Year 5	Year 6	Year 7	Year 8			
Labor												
Material												
Annual Total Maint.												
Equivalent Annual Cost												
Total Equiv. Annual Costs												
Net Present Value												

Figure 4

In order to fill in the "Equivalent Annual Cost" row, you will need to compute the present values (PV) of the Annual Total Maintenance Costs. Using the given formula for PV, enter a function in your calculator which computes PV as a function of time for this furniture. (You should assume the same 7% Discount Rate over the lifespan of the furniture.)

What function did you enter?

Using your calculator and the function you entered above, complete the chart in figure 4.

Based on these computations, explain which option seems more cost effective and why?

TASK 3: AVERAGE ANNUAL COSTS

Our previous method still does not take into account the different lifespans of the options.

Compute the *Average Annual Cost* for each option by dividing the Net Present Value by the lifespan of that option.

Based on this computation, explain which option seems more cost effective and why?

TASK 4: GRAPHICAL REPRESENTATION OF PRESENT VALUE

To better understand the effects of the present value computation, we can use a graphical representation. We will compare the effects of a $1500 annual fee versus a $1200 annual fee. If we use the function from the beginning of Task 2 to plot the Equivalent Annual Cost as a function of time, we get the following graph (shown here extended over 11 years).

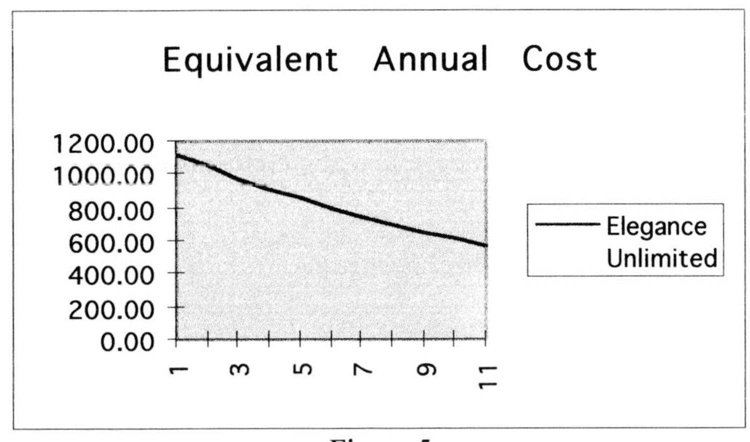

Figure 5

When analyzing the growth (or, as in this case, the decay) of a function, two models are most prevalent.

In the first (a *linear model*), the function grows (or decays) by a constant amount each time period. Specifically, for a function $f(t)$,the difference, $f(t+1)-f(t)$, would remain constant for different values of t.

For the second model (an *exponential model*), the function grows (or decays) by a fixed percent of its current value. In this case, it is the quotient, $f(t+1)/f(t)$, which remains constant for different values of t.

Examine the change in the Equivalent Annual Cost from year to year and explain which of these two models of growth best describes it.

On the same set of axes, for each brand of furniture, plot the Equivalent Annual Cost of the maintenance fee over 20 years.

Examine the ratio of the Equivalent Annual Costs of the two brands of furniture from year to year by filling in the table below. How do the Equivalent Annual Costs compare over time?)

	Year 1	Year 2	Year 3	Year 4	Year 5
Elegance Unlimited EAC					
Chairs R Us EAC					
Ratio Elegance/ ChairsRUs					

Write an expression for the ratio of the Equivalent Annual Costs as a function of time by dividing the function for the Present Value of Elegance Unlimited Furniture (from the beginning of Task 2) by the function for the Present Value of the Chairs R Us Furniture (from the middle of Task 2). Write the resulting expression in simplest form.

TASK 5: REPLACEMENT AND RESIDUAL VALUE

Although Task 3 considered average annual costs, there is still a problem in that the average annual costs will be paid over different lifespans. This needs to be equalized in order to make the comparisons more valid.

To do this, examine all the costs over an eleven year period. This has already been done for the Elegance Unlimited furniture, but our computation for Chairs R Us needs to be expanded to eleven years. Task 4 extended the maintenance costs, but there are still the initial costs to consider. Were the company to elect Chairs R Us as a vendor, we would need to replace the furniture after year 8. So, in the year 8 column, an additional expense of $11,000 needs to be added. At the end of eleven years, however, they will be left with a set of furniture which is still reasonably new (3 years old out of an 8 year lifespan). To account for this, the computations need to incorporate the notion of *residual value*. After year 11, assume that the furniture still retains 62.5% (5/8) of its initial value (The Internal Revenue Service is fond of straight-line (linear) depreciation). Therefore, in year 11, the account is *credited* with 62.5% of $11,000.

To extend our computation for Chairs R Us furniture to 11 years, we will use the chart shown in figure 6. Copy your initial costs and annual maintenance costs from figure 4 to this chart, extending the annual maintenance costs to 11 years.

Chairs 'R Us Furniture												
Discount Rate	7%											
	Initial Costs											
Purchase Price												
Installation/Delivery												
Total Initial Cost												
Annual Maintenance		Year 1	Year 2	Year 3	Year 4	Year 5	Year 6	Year 7	Year 8	Year 9	Year 10	Year 11
Labor												
Material												
Replace/Residual Value												
Annual Total												
Equivalent Annual Cost												
Total Equiv. Annual Costs												
Net Present Value												
Average Annual Cost												

Figure 6

This chart has an additional row labelled "Residual/Replacement Value". Enter the \$11,000 replacement cost in this row in year 8 and the residual value (as a negative number) in year 11. These entries should be used to recompute new annual costs and the annual costs should be used to recompute the Present Values and the Average Annual Cost over 11 years.

Based on the computations you have done, what would be your recommendation to a corporate manager concerning the two furniture options? Write a paragraph explaining your recommendation.

GLOSSARY:

Annual Percentage Rate (APR): Since interest is paid by different methods (eg. annually, quarterly, monthly, etc.), it is sometimes difficult to compare different payment options. For example, $100 invested at 6% compounded quarterly will be worth $106.14 after one year (because of compounding). In this case, Annual Percentage Rate is defined to be 6.14%. In general, the annual percentage rate is the equivalent percentage rate if interest were paid annually.

EXTENSION:

Add another row to your chart to keep a running total of the Equivalent Annual Costs. On one set of axes, graph this Total Equivalent Annual Cost versus time for both brands of furniture. What does the graph show you?

MATHEMATICS LABORATORY INVESTIGATION

LOGIC GATES

Topic: **NEGATIONS, CONJUNCTIONS, AND DISJUNCTIONS**
Prerequisite knowledge: *None*

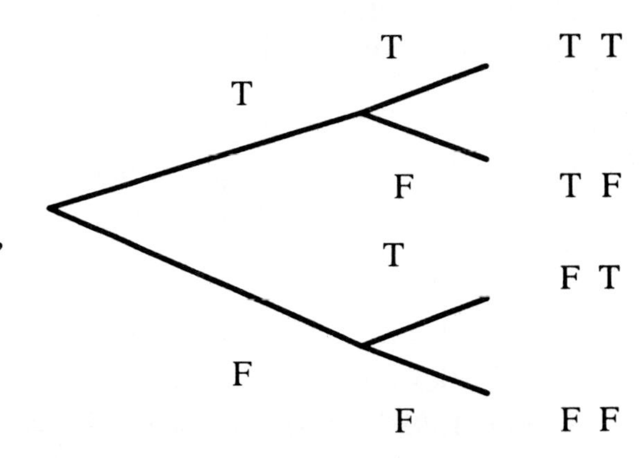

Equipment: Breadboard and wires, 7404, 7408, 7411, and 7432 logic gate chips, and a graphics calculator.

I. INTRODUCTION:

"New York City is a city in the State of New York" is a statement that happens to be a true statement. Its truth value is true. "7 + 3 = 12" is a statement that happens to be a false statement. Its truth value is false. These two statements are examples of simple statements. In this laboratory we will investigate compound statements, the negation of a statement, and their truth values.

"New York City is **not** a city in the State of New York" is a negation of the statement "New York City is a city in the State of New York". The following are examples of compound statements: "New York City is a city in the State of New York **and** 7 + 3 = 12"; "New York City is a city in the State of New York **or** 7 + 3 = 12"; "**if** New York City is a city in the State of New York, **then** 7 + 3 = 12".

The words **not, and, or,** as well as the combination of words **if - then** are used to form negations of statements and compound statements. Although we are studying negations and compound statements within the context of a mathematics course, what you will learn is universally applied to all scientific and engineering disciplines as well as non-scientific and non-engineering disciplines such as history, english, economics and sociology. In order to emphasize this point, these key words will be bold-faced throughout this laboratory investigation whenever they are used to form negations or compound statements.

We will investigate negations and compound statements by means of logical gates and graphics calculators. Logical gates are used extensively in electrical engineering, computer hardware engineering, and computer software design. Graphics calculators usually have logic capability **and** allow you to simulate relatively simple logic circuits.

II. STATEMENT VARIABLES

In algebra letters such as x, C, and R are used to represent numbers. $2x + 1 < 9$ is an algebraic inequality in which x represents any number less than 4. Hence x is called a <u>number variable</u>. In the formula $C = 2\pi R$, C and R are letters which represent numbers associated with the circumference and the radius of a circle. Hence C and R are <u>number variables</u>. π is a Greek letter which represents the <u>constant number</u> 3.14159265... Hence, π is **not** a number variable.

In logic, we also have a need for variables. However, the variables do not represent numbers. In logic, variables are used to represent statements. For instance, the statement "New York City is a city in the State of New York" could be represented by the <u>statement variable</u> P **and** the statement "7 + 3 = 12" could be represented by the statement <u>variable</u> Q. The compound statements "New York City is a city in the State of New York **and** 7 + 3 = 12"; "New York City is a city in the State of New York **or** 7 + 3 = 12"; "**if** New York City is a city in the State of New York, **then** 7 + 3 = 12" could be represented concisely by the variable statements P **and** Q, P **or** Q, and **if** P, **then** Q.

Some authors use upper case letters for statement variables, others use lower case letters. Upper case letters will be used throughout this laboratory investigation.

III. NEGATIONS

"New York City is a city in the State of New York" is a true statement. Its negation is the statement "New York City is **not** a city in the State of New York", a false statement. "7 + 3 = 12" is a false statement. Its negation is the statement "$7 + 3 \neq 12$", a true statement. **A statement and its negation always have opposite truth values.**

TASK1: <u>Forming negations</u>.

Write the negations of the following statements.

1. I am enrolled in an economics class.

2. $5 + 3 = 8$

3. All artists are music lovers.

4. $5 + 3 \geq 8$

5. A triangle has exactly two sides.

6. $5 + 3 < 8$

IV. NEGATION NOTATION AND TRUTH TABLE

If the statement "New York City is a city in the State of New York" is represented by the statement variable P, **then** its negation "New York City is **not** a city in the State of New York"is represented in most mathematics books as **~P**. However, you will find that on many graphics calculators the negation of P is represented as **not P and** in most electrical engineering and computer hardware books the negation of P is represented as **-P**. We need to be comfortable with all three notations.

TASK 2: Using negation notation.

Return to TASK 1 and assign a different statement variable to each of the 6 statements that were given. Then assign the negation notation to each of the negations which you have written, using the mathematics notation for statements 1 and 2, the graphics calculator notation for statements 3 and 4, and the engineering notation for statements 5 and 6.

TASK 3: Creating a negation truth table.

A truth table is a very useful device for determining the truth value of a complex compound statement. The truth table for a negation is very simple. Given a statement represented by a statement variable R, R can have only two possible truth values, true or false. Hence the first column of the truth table lists R and its two possible truth values.

R
T
F

The second column of the truth table lists the notation for the negation of R and the truth values for the negation of R that correspond to the truth values of R in the first column of the table. Fill in the following table and check the results with your teacher.

R	**~R**
T	
F	

Now create two truth tables for the negation of R at the bottom of the page, one using the **not R** notation and one using the **-R** notation.

TASK 4: Using binary representation of truth values.

Since a statement has only two possible truth values, true or false, most graphics calculators and computer languages assign the binary values of 1 and 0 to the truth values true and false. 1 represents true and 0 represents false.

Fill in the 3 truth tables for the negation of R using binary truth values

R	~R		R	not R		R	-R

IV. NEGATIONS ON A GRAPHICS CALCULATOR AND BREADBOARD

TASK 5: Negations on the graphics calculator.

With the help of your instructor, locate the logic menu on your graphics calculator. The calculator can be used in two differeent ways.

First, enter **not** 1 (not true) and the calculator should return 0 (false). Enter **not** 0 (not false) and the calculator should return 1 (true).

Second, assign the value of 1 to the variable A. What valuc is returned by the calculator when you now enter **not** A?

Now assign the value of 0 to the variable A. What value is returned by your calculator when you enter **not** A?

Complete the truth table for the variable A.

A	not A

TASK 6: <u>Negations on a breadboard.</u>

The electronic equivalent of a negation is an inverter. An inverter is a single-input, single-output device (a single-input *function*). The truth value of the input signal is converted into the opposite truth value which is then transmitted through the output of the device. The symbol for an inverter is portrayed here. The signal enters from the left side of the device **and** exits through the bubble.

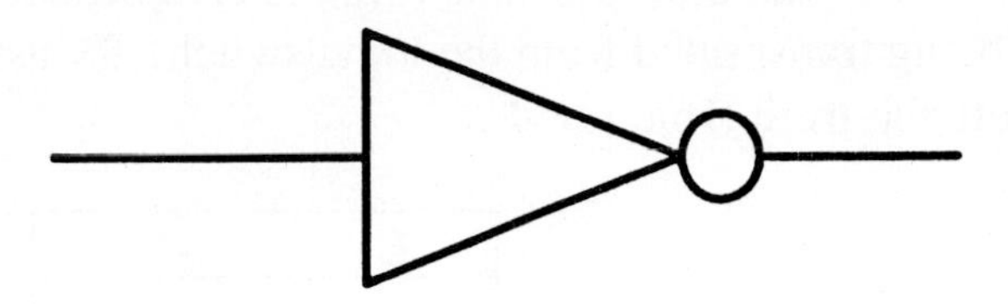

The 7404 chip has 6 inverters built on the chip **and** is called hex inverters in Appendix A. Locate the schematic of the '04 chip.

Notice the dimple on the left side of the schematic. This dimple corresponds to the dimple on the '04 chip. In order to properly orient the chip, the dimple must be to the left. With the chip properly oriented, the pin numbers on the schematic correspond to the pins on the chip.

Your instructor will assist you in connecting an inverter. However, you should first develop a schematic showing the input pin and the output pin of the inverter you choose. Suppose we were to select the inverter whose input is pin 11 and whose output is pin 10. Further suppose that the input was coming from a logic switch on the breadboard **and** that the output was going to a lamp monitor on the breadboard. The schematic would be drawn as follows.

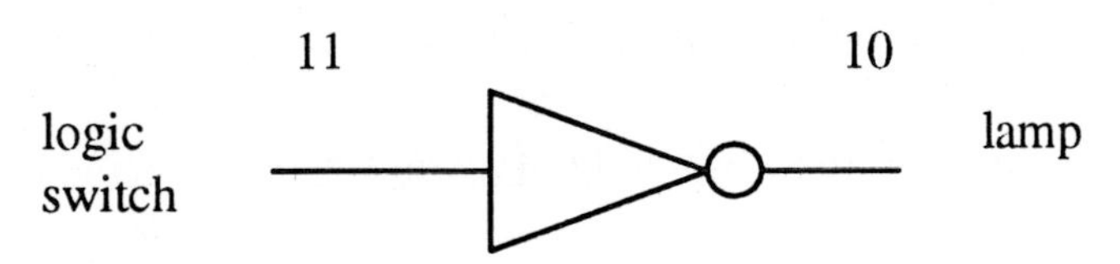

Choose a different inverter and draw the corresponding schematic below.

You are now ready to connect the inverter. With the power source off, connect the ground to pin 7. Connect the volt source to pin 14. Connect a logic switch to your input pin. Connect your output pin to a lamp monitor. Put the power switch on.

When the logic switch is in the up position a 1 signal or a true signal is transmitted to the inverter. When the logic switch is in the down position a 0 signal or a false signal is transmitted to the inverter.

When the inverter transmits a 1 (true) signal, the lamp is lit. When the inverter transmits a 0 (false) signal, the lamp is **not** lit.

Let the statement variable L represent a statement whose truth value is the signal being transmitted from the logic switch. By using the logic switch in both positions complete the truth table for **-L**.

L	**-L**

V. CONJUNCTIONS

The compound statement "New York City is a city in the State of New York **and** 7 + 3 = 12" is an example of a conjunction. **When two or more simple statements are joined by the word <u>and</u>, the resulting compound statement is called a conjunction.**

Any compound statement has a truth value. The truth value of a compund statement is dependent upon the type of the compound statement and the truth values of the simple statements used in the compound statement. Truth tables display the relationship between the truth value of a compound statement and the truth values of the simple statements used in the compound statement.

TASK 7: <u>Deriving the truth table of a conjunction</u>.

We are going to derive truth tables for conjunctions by using logical gates called <u>AND Gates</u>. The simplest possible conjunction is a compound statement in which <u>exactly</u> two simple statements are joined by the word <u>and</u>. "New York City is a city in the State of New York **and** 7 + 3 = 12" is an example of such a conjunction.

Let P and Q be statement variables that represent two simple statements. Then **P and Q** represents a conjunction. **P and Q** is the notation used on most graphics calculators. **P∧Q** is the notation used in most mathematics books. The multiplication notation **PQ** is used in most engineering books.

Since the statement variable P has two possible truth values (true or false) **and** the statement variable Q also has two possible truth values (true or false), **then** the conjunction P **and** Q has $2 \times 2 = 4$ different possible combinations of pairs of truth values for the simple statements. Therefore, the truth table for P **and** Q will have to begin as follows on the next page.

P	Q
T	T
T	F
F	T
F	F

(See the 2-level tree diagram on the first page of this laboratory investigation.)

We now need to insert a column in which to record the truth value of P **and** Q for each of the four possible pairs of truth values for P and Q. <u>Do not attempt to fill in the truth table at this time</u>.

P	Q	P and Q
T	T	
T	F	
F	T	
F	F	

In Appendix A locate the Quad 2-input AND Gates chip. This is the 7408 chip. Its name is derived from the fact that the chip includes 4 AND Gates and each AND Gate is a 2- input, 1-output device (a 2-input *function*).

Mount a 7408 chip on the breadboard being sure that the dimple is to the left. Connect the chip to ground and to the power supply. If the AND Gate in the lower left corner were chosen, each of the input pins 1 and 2 would have to be connected to a different logic switch. The output pin 3 would be connected to a lamp monitor. The schematic would be as follows.

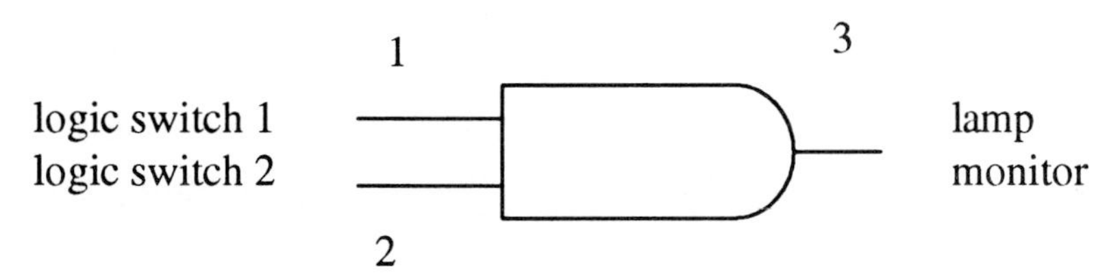

Choose a different AND Gate and draw a similar schematic below.

Now connect the AND Gate which you have chosen and by experimentation derive <u>only the left-most truth table</u> at this time. The other two tables will be completed shortly.

P	Q	PQ		P	Q	P and Q		P	Q	$P \wedge Q$	
1	1			1	1			T	T		
1	0			1	0			T	F		
0	1			0	1			F	T		
0	0			0	0			F	F		

Verify the result by using the **and** option on the logic menu of your graphics calculator and now complete the middle truth table. (Hint: The first row of the table can be entered simply as 1 **and** 1. There is no need to assign truth values to P and Q.)

Now complete the right-most truth table.

Based upon your results, determine and report the truth value of the conjunction "New York City is a city in the State of New York **and** 7 + 3 = 12".

In Appendix A locate the 7411 chip. The 7411 triple 3-input AND Gates has three 3-input AND Gates. A 3-input AND Gate is a triple-input, single-output device (a triple-input *function*). Since the Gate has 3 inputs, there are $2 \times 2 \times 2 = 8$ possible combinations of input triples. Draw a 3-level tree diagram to derive the 8 possible combinations. See the 2-level tree diagram on the first page of this laboratory investigation.

Mount a 7411 chip on the breadboard connecting the chip to ground and to the power supply. Select a 3-input And Gate and complete the following schematic.

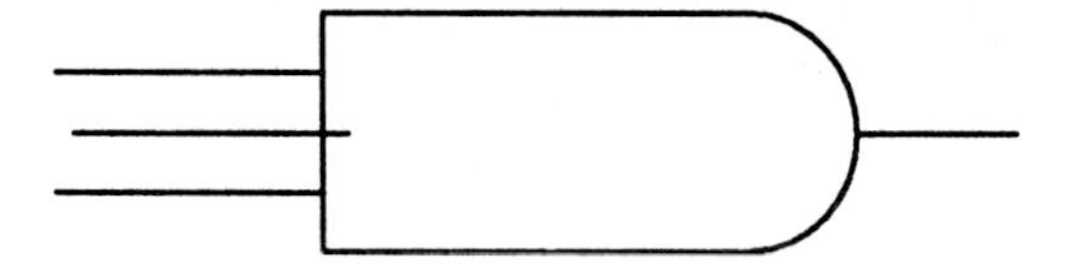

Connect the AND Gate and by experimentation complete only at this time the left-most truth table for the conjunction **PQR.**

P	Q	R	PQR		P	Q	R	P and Q and R		P	Q	R	P∧Q∧R	
1	1	1			1	1	1			T	T	T		
1	1	0			1	1	0			T	T	F		
1	0	1			1	0	1			T	F	T		
1	0	0			1	0	0			T	F	F		
0	1	1			0	1	1			F	T	T		
0	1	0			0	1	0			F	T	F		
0	0	1			0	0	1			F	F	T		
0	0	0			0	0	0			F	F	F		

Verify your results on a graphics calculator and now complete the center truth table for the conjunction **P and Q and R.**

Now complete the right-most truth table for **P∧Q∧R.**

Having now studied the conjunction of two statements and the conjunction of three statements, fill in the blanks for the following statements.

A conjunction is a true statement when ______________ of the simple statements are ______________ statements.

A conjunction is a false statement if ________________ of the simple statements is a __________________ statement.

VI. DISJUNCTIONS

The compound statement "New York City is a city in the State of New York **or** 7 + 3 = 12" is an example of a disjunction. **When two or more simple statements are joined by the word or, the resulting compound statement is called a disjunction.**

TASK 8: Deriving the truth table of a disjunction

We are going to derive truth tables for disjunctions by using OR Gates. The simplest possible disjunction is a compound statement in which exactly two simple statements are joined by the word or. "New York City is a city in the State of New York **or** 7 + 3 = 12" is an example of such a disjunction.

Let P and Q be logical variables that represent two simple statements. Then **P or Q** represents a disjunction. **P or Q** is the notation used on most graphics calculators. **P∨Q** is the notation used in most mathematics books. The addition notation **P + Q** is used in most engineering books.

The truth table for **P or Q** is as follows. Do not fill it in at this time.

P	Q	P or Q
T	T	
T	F	
F	T	
F	F	

In Appendix A locate the Quad 2-input OR Gates chip. This is the 7432 Chip. Its name is derived from the fact that the chip includes 4 OR Gates and each OR Gate is a 2- input, 1-output device (a 2-input *function*).

Mount a 7432 chip on the breadboard connecting the chip to ground and to the power supply. Choose an OR Gate and complete the following schematic.

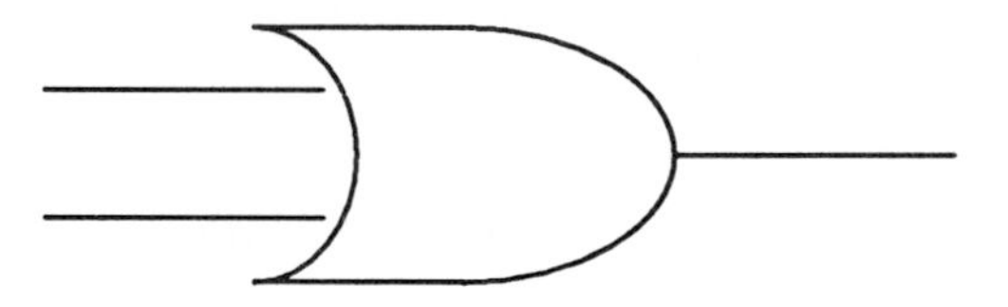

Now connect the OR Gate which you have chosen.

By experimentation derive only the left-most truth table for P + Q.

P	Q	P + Q		P	Q	P or Q		P	Q	P ∨ Q
1	1			1	1			1	1	
1	0			1	0			1	0	
0	1			0	1			0	1	
0	0			0	0			0	0	

Verify the result by using the **or** option on the logic menu of your graphics calculator and now complete the central truth table for **P or Q**.

Now fill in the right-most table for **P ∨ Q**.

Based upon your results, determine and record the truth value of the disjunction "New York City is a city in the State of New York **or** 7 + 3 = 12".

There is no chip for 3-input OR Gates in Appendix A. By using the logic capability of your graphics calculator, derive and fill in the truth table for **P or Q or R.**

P	Q	R	P or Q or R
T	T	T	
T	T	F	
T	F	T	
T	F	F	
F	T	T	
F	T	F	
F	F	T	
F	F	F	

The schematic for a 3-input OR Gate is portrayed here.

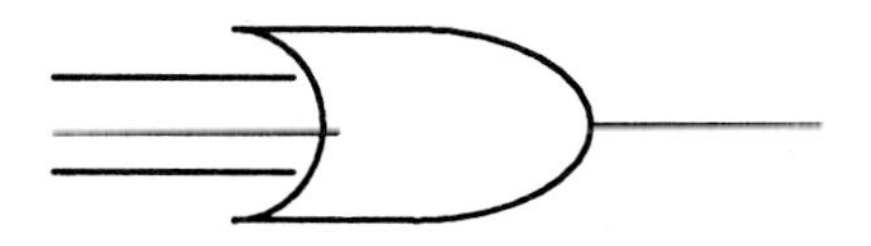

Having now studied the disjunction of two statements and the disjunction of three statements, fill in the blanks for the following statements.

A disjunction is a true statement when _____________ of the simple statements is a _____________ statement.

A disjunction is a false statement if _______________ of the simple statements are __________________ statements.

VII. HOMEWORK

TASK 1: Select a page from a book, a magazine, or a newspaper and highlight or underline all conjunctions, disjunctions, and conditionals.

TASK 2: Based upon the results from this laboratory complete the following truth tables. Note that with an input of 4 simple statements, $2 \times 2 \times 2 \times 2 = 16$ 4-tuple inputs are possible.

P	**Q**	**R**	**S**	**P∧Q∧R∧S**	
1	1	1	1		
1	1	1	0		
1	1	0	1		
1	1	0	0		
1	0	1	1		
1	0	1	0		
1	0	0	1		
1	0	0	0		
0	1	1	1		
0	1	1	0		
0	1	0	1		
0	1	0	0		
0	0	1	1		
0	0	1	0		
0	0	0	1		
0	0	0	0		

P	**Q**	**R**	**S**	**P∨Q∨R∨S**
1	1	1	1	
1	1	1	0	
1	1	0	1	
1	1	0	0	
1	0	1	1	
1	0	1	0	
1	0	0	1	
1	0	0	0	
0	1	1	1	
0	1	1	0	
0	1	0	1	
0	1	0	0	
0	0	1	1	
0	0	1	0	
0	0	0	1	
0	0	0	0	

TASK 3: Determine and record the truth value of the following statements.

a) 2 is an even integer **and** 8 is an odd integer.

b) 2 is an even integer **or** 8 is an odd integer.

c) $2 < 3$ **and** 3 is a positive integer **and** New York City is in New York State.

d) 2 is an odd integer **or** 8 is an **odd** integer or New York City is in Maine.

TASK 4: Complete the following task only if you are familiar with functions, domains of functions, and ranges of functions.

In the following consider the inputs and outputs to the devices to be the binary numbers 1 and 0.

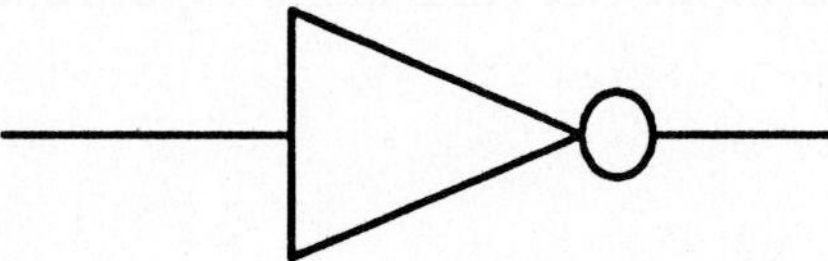

The inverter is a single-input function. List the domain and the range of the inverter.

Domain = { } Range = { }

On a number line graph the domain of the inverter.

On a number line graph the range of the inverter.

In two dimensions graph the inverter function.

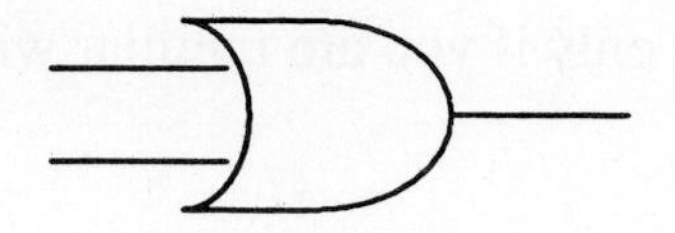

The 2-input OR Gate is a 2-input function. List the domain and range of the 2-input **OR** Gate.

In two dimensions graph the domain of the 2-input OR Gate.

On a number line graph the range of the 2-input OR Gate.

Using 3-dimensional graphing, graph the 2-input OR Gate function.

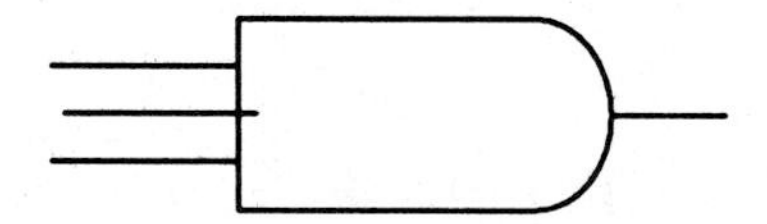

The 3-input AND Gate is a 3-input function. List the domain and range of the 3-input AND Gate.

Using 3-dimensional graphing, graph the domain of the 3-input And Gate.

On a number line graph the range of the 3-input AND Gate.

APPENDIX A

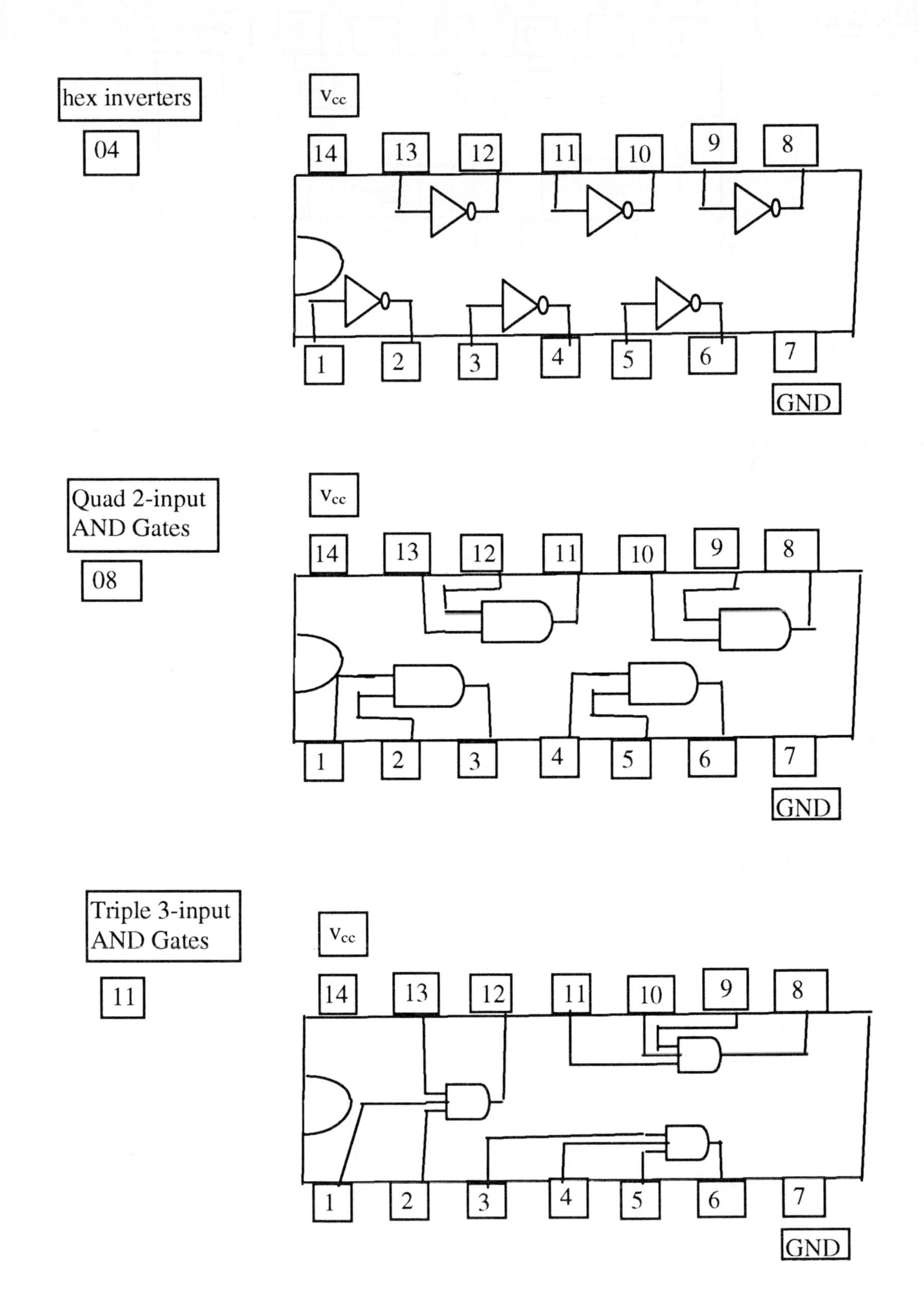
hex inverters
04
Vcc
14
13
12
11
10
9
8
1
2
3
4
5
6
7
GND
Quad 2-input AND Gates
08
Vcc
14
13
12
11
10
9
8
1
2
3
4
5
6
7
GND
Triple 3-input AND Gates
11
Vcc
14
13
12
11
10
9
8
1
2
3
4
5
6
7
GND

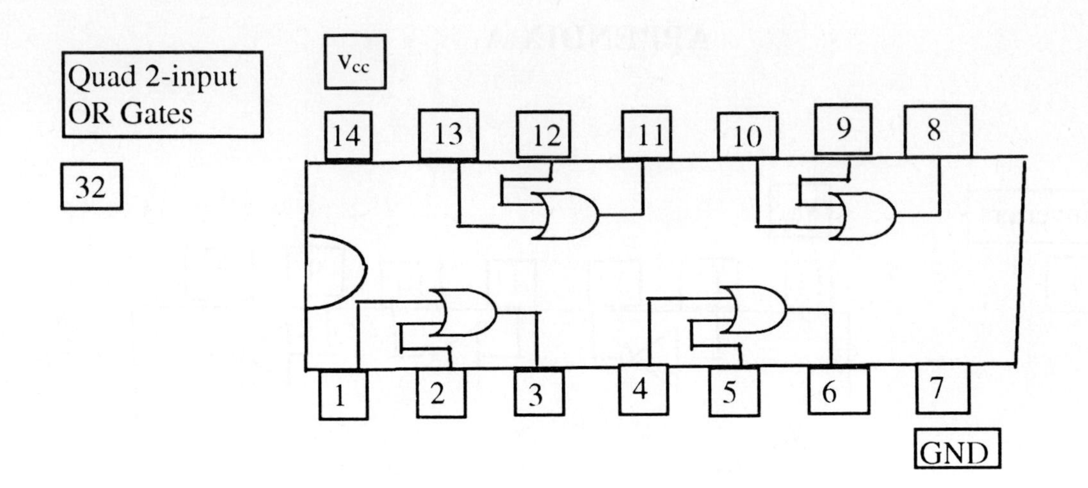
Quad 2-input
OR Gates
32
Vcc
14
13
12
11
10
9
8
1
2
3
4
5
6
7
GND

MATHEMATICS LABORATORY INVESTIGATION

MEDICATION DOSING

Topic : **A FUNCTION GENERATED BY CONVOLUTION SUMMATION, MODELING**

Prerequisite knowledge : *Convolution Summation, Half-life, Nonlinear Regression, Unit Conversions, Computing Interest II Investigation*

Equipment: Graphics calculator or computer software applications, graph paper, transparent grid.

I. INTRODUCTION :

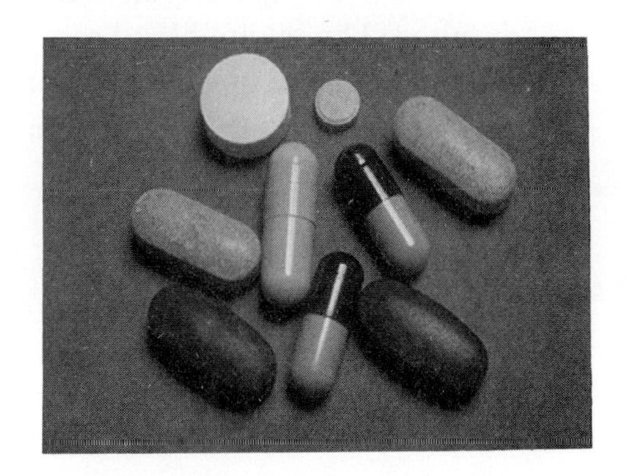

Certain over the counter drugs such as aspirin are taken periodically, every four hours, for example. The drugs enter the bloodstream, reach their maximum concentration in the bloodstream, and, for all practical purposes, dissipate completely before the next scheduled dose is taken. The second dose of the drugs does not build upon the first dose of the drugs.

Other drugs, often prescribed drugs, such as antihistamine drugs or anticonvulsant drugs, enter the bloodstream, reach a maximum concentration in the bloodstream, and only partially dissipate before the next scheduled dose. When the next scheduled dose is taken, the total concentration of these drugs in the bloodstream is the sum of the residual concentration from the previous dose(s) as well as the concentration from the present dose.

Drugs such as anticonvulsants are designed to be be effective when their concentrations in the bloodstream are within a prescribed range. Initial doses are repeated and/or adjusted until the concentrations are within the prescribed range.

Anticonvulsants and many other drugs have half-lifes. By applying our knowledge of half-lifes we can ***model*** the concentration of the drugs in the bloodstream for a prescribed dosing routine. We can then create different ***models*** for different dosing routines to eventually arrive at an effective dosing schedule.

Your study of the **Convolution Summation** in the Computing Interest Laboratory Investigations provides you with a mathematical tool that is useful in keeping track of the residual concentrations of the drugs from previous doses as well as the most recent dose.

II. THE PROBLEM

In order to control his seizures, John is required to begin taking a new anticonvulsant. The anticonvulsant's half-life varies from 6 hours to 16 hours depending upon various conditions, such as the presence of other anticonvulsant drugs in the bloodstream. The anticonvulsant is taken orally and ingested into the bloodstream through the stomach and intestines. Approximately 50% of the anticonvulsant is removed from the bloodstream during its first pass through the liver. Of the remaining 50%, approximately 90% of the the anticonvulsant is bound to plasma protein in the blood, the other 10% being unbound in the blood. The drug is given in pills, each containing 250 mg of the anticonvulsant.

The effective level of the anticonvulsant in the bloodstream can be monitored by taking blood samples. It is the anticonvulsant that is bound to plasma protein that is measured in a blood sample. The desired range of the concentration of the bound anticonvulsant in the bloodstream is 50 - 100 micrograms / milliliter.

John weighs 157.5 lbs. The average volume of blood in a person is approximately 70 milliliters / kilogram. (1 lb = 0.4536 kg)

John's initial dosing schedule is 2 anticonvulsant pills at 6:00 AM, 2:00 PM, and 10:00 PM. This dosing schedule is to be repeated daily until the doctor changes it.

The problem is to mathematically model the concentration of anticonvulsant in John's bloodstream in μg / ml .

III. LET'S TALK IT UP

Get together with your lab partners at the beginning of the lab or preferably before the lab period to discuss the problem. Arriving at an actual solution might be difficult, but give it a try. Some of you may have different approaches. Try more than one. See where you agree and where you disagree.

What approach(es) did you settle on?

Were you able to estimate the range of anticonvulsant in John's bloodstream?

What assumptions did you make in arriving at your solution?

Do you think that your solution is accurate to the nearest μg / ml ?

Be prepared to make a short presentation to the class.

III. A GRAPHICAL APPROACH TO THE PROBLEM

TASK 1 : Preparing a half-life graph.

Assume a half-life of 8 hours for the anticonvulsant. On a grid transparency use your knowledge of half-life to find several points on the graph depicting the concentration, in μg/ml, of the bound anticonvulsant in the bloodstream for **one** pill for a period of time of 6 half-lifes. **Increment the time axis in 2-hour time intervals**. Carefully draw in a smooth curve. Properly label and increment the vertical and horizontal axes. Let's call this the **Unit Concentration Function** (unit meaning one pill).

TASK 2 : Obtaining the equation of the Unit Concentration Function.

Obtain an equation of the Unit Concentration Function by using your knowledge of half-life functions or by using non-linear regression on a graphics calculator or a computer software application.

$U(t) =$

TASK 3 : Graphing the Dosage Function.

On graph paper, graph the **Dosage Function.** The horizontal axis should be the same as in the Unit Concentration Function. The vertical axis should correspond to the number of pills taken.

TASK 4 : Using Convolution Summation.

RECALL: $(D * U)(10) = \sum_{t=0}^{10} D(t) \cdot U(10 - t)$, is the convolution of the Dosage and Unit Concentration Functions at time 10 hrs. This convolution represents the concentration of bound anticonvulsant in the bloodstream at time 10 hrs.

Graphically, the Unit Concentration Function is layed over the Dosage Function and shifted 10 hrs to the right and then reflected through the vertical line t = 10. Multiplications on or to the left of line t = 10 are performed and the products are summed, yielding the concentration of the bound anticonvulsant at t = 10 hrs. Concentrations at other times are calculated similarly.

Graphically determine the bound anticonvulsant concentrations for the first 24 hours of dosing. Record your results in the chart to the nearest μg / ml .

time, hrs.	0	1	2	3	4	5	6	7
concentration, μg / ml								

time, hrs.	8	9	10	11	12	13	14	15
concentration, μg / ml								

time, hrs.	16	17	18	19	20	21	22	23
concentration, μg / ml								

time, hrs.	24
concentration, μg / ml	

TASK 5 : Graphing the Bound Anticovulsant Concentration Function.

The information obtained in the chart in task 4 provides 25 points in the graph of the Bound Anticonvulsant Concentration Function for the particular dosing schedule and half-life which we used.. Graph these points on a piece of graph paper. Carefully label and increment each axis. Draw a smooth curve through the 25 points.

What are your observations about the graph?

What are your predictions for the graph if we were to continue beyond 24 hours?

Using a different colored pen or pencil extend the graph through the second day according to your predictions for the graph.

TASK 6: Completing the Graph of the Bound Anticonvulsant Function

The following data was derived by means of a program which was actually written for a programmable calculator. The program utilized the symbolic representation of convolution summation and actually calculated the summation on an hourly basis.

Hours	µg/ml	Hours	µg/ml	Hours	µg/ml
0	45	8	67.5	16	78.75
1	41.3	9	61.9	17	72.2
2	37.8	10	56.8	18	66.2
3	34.7	11	52	19	60.7
4	31.8	12	47.7	20	55.7
5	29.2	13	43.8	21	51.1
6	26.8	14	40.1	22	46.8
7	24.5	15	36.8	23	42.9

Hours	µg/ml	Hours	µg/ml	Hours	µg/ml
24	84.4	32	87.2	40	88.6
25	77.4	33	80	41	81.2
26	71	34	73.3	42	74.5
27	65.1	35	67.2	43	68.3
28	59.7	36	61.7	44	62.6
29	54.7	37	56.5	45	57.4
30	50.2	38	51.8	46	52.7
31	46	39	47.5	47	48.3

Hours	µg/ml	Hours	µg/ml	Hours	µg/ml
48	89.3	56	89.6	64	89.8
49	81.9	57	82.2	65	82.4
50	75.1	58	75.4	66	75.5
51	68.9	59	69.1	67	69.3
52	63.1	60	63.4	68	63.5
53	57.9	61	58.1	69	58.2
54	53.1	62	53.3	70	53.4
55	48.7	63	48.9	71	49
				72	89.9

To plot the data on an hourly basis would be tedious and unnecessary. Plot the data on a two-hour basis ($t = 0,2,4,6,\ldots,72$) and carefully draw in a continuous curve through the data points. Carefully label and increment each axis.

Compare the graph with the graph that you predicted in TASK 5 and discuss the similarities and differences in the two graphs.

By studying the graph and the data presented in tabular form, do we need to continue the Convolution Summation beyond 72 hours to determine the behavior of the Bound Anticonvulsant Concentration Function? Defend your response.

By studying the graph and the data presented in tabular form, estimate the maximum and minimum concentrations of the bound anticonvulsant in the bloodstream.

Is there a need to adjust the dosing schedule? Defend your response. If a change is needed what change would you suggest?

TASK 7: More Modeling.

Suppose that the dosing schedule used has caused John to have an upset stomach. In order to alleviate the distress John's doctor has recommended that John take his pills at meal time. John normally eats at 7:00 AM, 12:00 PM, and 6:00 PM. Draw a new Dosage Function for a 24 hour period.

TASK 8: Completing the Graph of the Bound Anticonvulsant Function

The resulting Bound Anticonvulsant Concentration Function is again presented to you in tabular form on the next page. Plot the data points and draw a smooth curve through the data points.

By studying the graph and the data presented in tabular form, do we need to continue the Convolution Summation beyond 72 hours to determine the behavior of the Bound Anticonvulsant Concentration Function? Defend your response.

By studying the graph and the data presented in tabular form, estimate the maximum and minimum concentrations of the bound anticonvulsant in the bloodstream.

Is there a need to adjust the dosing schedule? Defend your response. If a change is needed what change would you suggest?

Hours	μg/ml
0	45.0
1	41.3
2	37.8
3	34.7
4	31.8
5	74.2
6	68.0
7	62.4

Hours	μg/ml
8	57.2
9	52.5
10	48.1
11	89.1
12	81.7
13	74.9
14	68.7
15	63.0

Hours	μg/ml
16	57.8
17	53.0
18	48.6
19	44.6
20	40.9
21	37.5
22	34.4
23	31.5

Hours	μg/ml
24	73.9
25	67.8
26	62.1
27	57.0
28	52.2
29	92.9
30	85.2
31	78.1

Hours	μg/ml
32	71.6
33	65.7
34	60.2
35	100.2
36	91.9
37	84.3
38	77.3
39	70.9

Hours	μg/ml
40	65.0
41	59.6
42	54.7
43	50.1
44	46.0
45	42.1
46	38.6
47	35.4

Hours	μg/ml
48	77.5
49	71.1
50	65.2
51	59.8
52	54.8
53	95.3
54	87.3
55	80.1

Hours	μg/ml
56	73.5
57	67.4
58	61.8
59	101.6
60	93.2
61	85.5
62	78.4
63	71.9

Hours	μg/ml
64	65.9
65	60.4
66	55.4
67	50.8
68	46.6
69	42.7
70	39.2
71	35.9
72	78.0

MATHEMATICS LABORATORY INVESTIGATION

MILLING MACHINE

Topic: **COORDINATE GEOMETRY, DOMAIN RESTRICTIONS**
Prerequisite knowledge: *Cartesian Coordinates, Equations of Lines and Circles, Definition of Function*

Equipment: flat wrench

I. INTRODUCTION

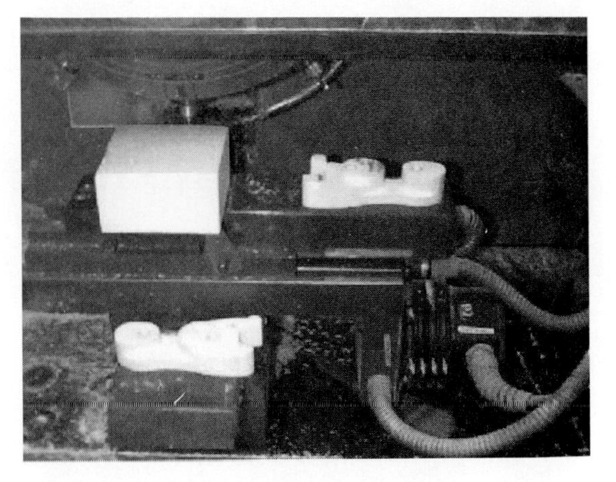

Milling is one of many processes used in manufacturing industries. When an object has to be cut precisely from a solid piece of stock material, it is milled. Common stock materials are wood, plastic and metal. Machine parts can be milled to very exact specifications. Molded plastic, glass or metal items are often cast in molds composed of a milled die. In another process, one pattern is created with a milling machine, several molds are made from that object and liquefied plastic, glass or metal is poured into the mold and left to harden.

II. MODERN MILLING MACHINE OPERATIONS

Older milling machines can be dangerous to operate. Modern computer driven milling machines require less human contact and are safer. A modern milling machine is pictured at right. An operator must instruct the computer how to direct the machine to cut out the desired object from the solid material. The operator's instructions must tell the machine where to begin and to end a cut, how deep to cut, and to cut in a straight line or circular arc. One method of instructing the computer driven machine is with codes. This, in effect, is writing a program to tell the machine how to cut from Point A to Point B and then from Point B to Point C and so on until the entire object has been cut from the stock material.

The geometry of creating machine instructions is similar to that used in many CAD (Computer Aided Design) programs. The process is the same for milling a pattern for your class ring or for creating an architectural drawing.

III. DESCRIBING THE MILLING PROCESS

The milling process can be described in many different ways. Different machines and different CAM (computer aided manufacturing) programs use different programming languages. This lab will be using a generic instruction language. Such machine-independent languages are referred to as *pseudocode*. For the particular pseudocode of this lab, assume that the milling machine is capable of making two kinds of cuts: straight lines and circular arcs. To see how the pseudocode works, consider the problem of milling a piece shaped like three quarters of a disk.

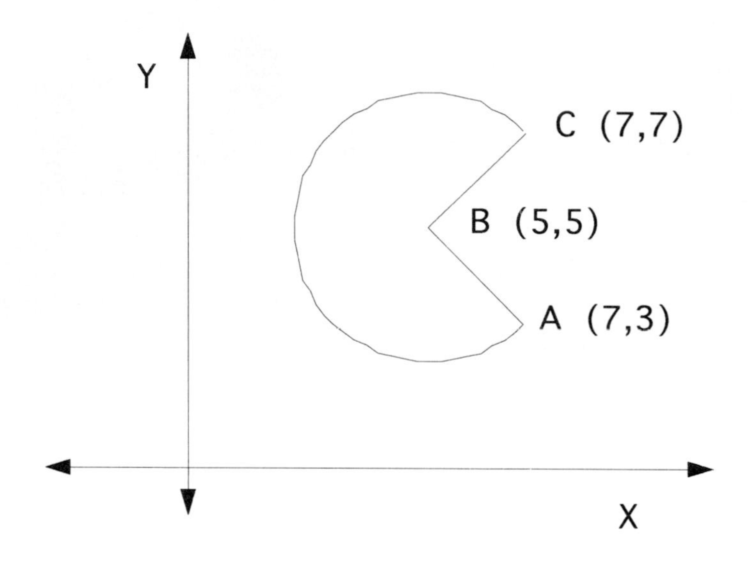

For this piece, the pseudocode could read:

Go from (7,3) to (5,5) cutting a straight line
Go from (5,5) to (7,7) cutting a straight line
Cut a circle of radius 2.83 centered at (5,5) counterclockwise
from (7,7) to (7,3)

Notice that there are three points where the various cuts meet. These three *cut points* are labelled in the picture by A, B, and C, while the pseudocode refers to them by their coordinates.

Telling the milling machine where to start and stop is easy. It's a little more complicated to describe the milling process by using algebraic equations. The equations describe the line segments and circular arcs outlining the piece to be milled. In the case of the first cut above, it is a straight line with slope -1 passing through the point (5,5). But only that portion of the line from x=5 to x=7 is needed. This is called a *domain restriction*. The use of domain restrictions gives what are called *piecewise defined functions*. The equation for such a function is given by:

$$y = -x + 10, (5 \le x \le 7)$$

For the circular cut, the situation is somewhat different. Recall that the equation of a circle, $x^2 + y^2 = r^2$, does not define a function. Rather, it requires two functions, one for the top half of the circle and one for the bottom half. In the case of this piece, the center of the circle is at (5,5) and the radius is $2\sqrt{2}$ or approximately 2.83. (The radius can be found by computing the distance between (5,5) and (7,7).) This gives the equations

$$y = 5 + \sqrt{8 - (x - 5)^2}, (2.17 \le x \le 7)$$

and

$$y = 5 - \sqrt{8 - (x - 5)^2}, (2.17 \le x \le 7)$$

Consequently, the milling can be summarized by the following table.

Cut	Pseudocode	Equation(s)
A to B	Go from (7,3) to (5,5) cutting a straight line	$y = -x + 10, (5 \le x \le 7)$
B to C	Go from (5,5) to (7,7) cutting a straight line	$y = x, (5 \le x \le 7)$
C to A	Cut a circle of radius 2.83 centered at (5,5) counter-clockwise from (7,7) to (7,3)	$y = 5 + \sqrt{8 - (x - 5)^2}, (2.17 \le x \le 7)$ and $y = 5 - \sqrt{8 - (x - 5)^2}, (2.17 \le x \le 7)$

The advantage of using algebra is that technology provides a way to check your equations so you can view your design before sending the pseudocode to the milling machine programmer. If your school has a modern milling machine, your code could be translated into its machine language and the shape would be milled from your design. Checking your design is important in the work place. You don't want to gain a reputation for errors which waste other employees' time and raw materials. Both of these are costly to the company, not just to your self esteem.

Try entering these equations in your graphing calculator and see if you get the original picture.

IV. DESIGNING THE MILLING PROCESS

TASK 1: Milling a Keyhole Shape

Consider the following sketch of a "keyhole shape" drawn in four different orientations on a grid background. Although the shape is the same in all four pictures, the various orientations require different pseudocode and can lead to difficulties in writing their equations. For example, if the intent is to describe the pictures entirely by functions, which of the four pictures can **not** be so described?

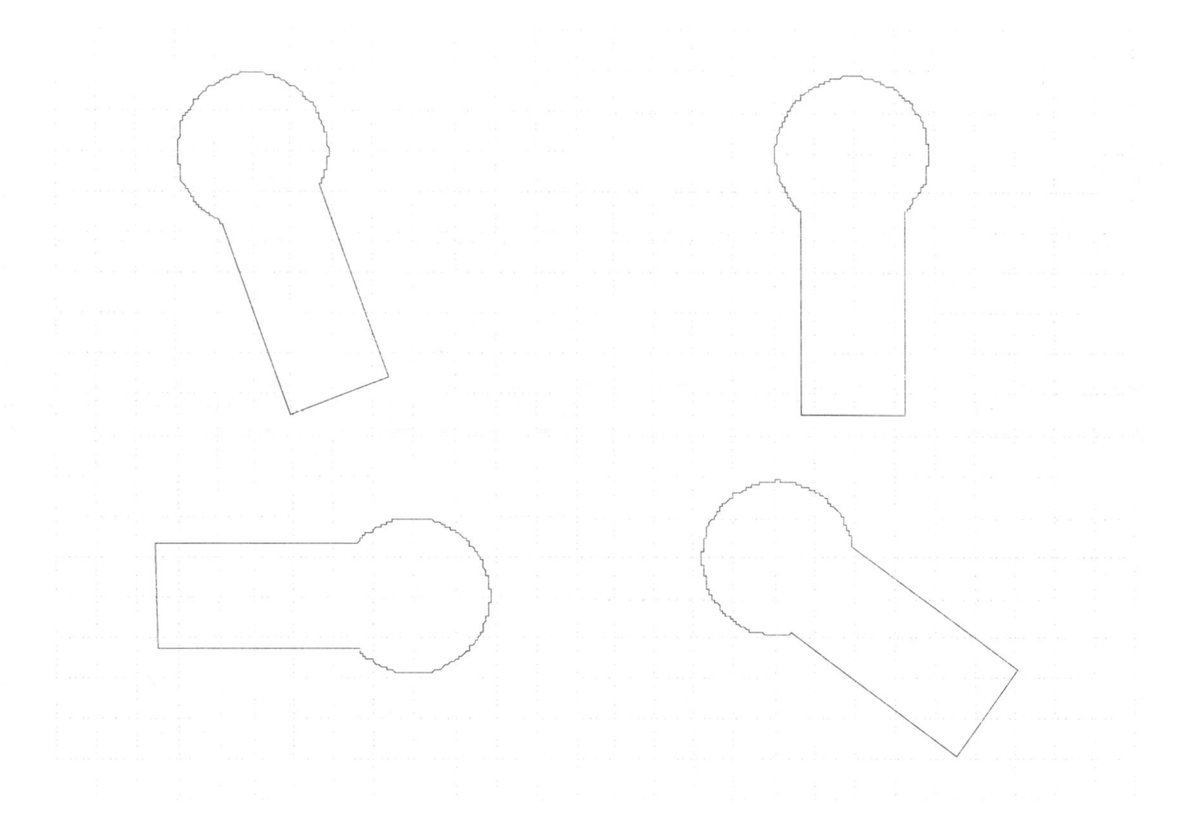

Choose one of the pictures which can be described entirely by functions. Label the cut points in the order that the milling machine will traverse them. Choose a point on the grid to serve as the origin and estimate the coordinates of your points to the nearest tenth of a grid unit. For circular cuts, estimate the center and radius of the circle to the same accuracy. Finally, complete **the first two columns** of the following chart (as in the previous example). (This chart may have more or fewer rows than you need. Feel free to adjust it if necessary.)

Cut	Pseudocode	Equation(s)

Once your pseudocode is complete, find the equations representing the individual cuts and fill in the third column of the chart.

Constructing an object often involves a large expense in materials and employee time/salary. It is almost always useful to check a design before investing time and materials into actually creating a physical object. Using a graphing calculator or computer graphing software, graph the equations listed in the third column and check to see that you reconstruct your original picture. (If not, correct the equations so you do.) (See the Appendix for details about using the graphing calculator.)

Check your work in this task against another group which chose a different picture. Did the choice of picture yield any significant difference in work necessary to fill in the chart?

TASK 2: Milling a Wrench

To mill a flat wrench, position a sample on a piece of graph paper such as the one included below. Since you will be checking your model on a graphing calculator (or similar computer software) and will therefore be using algebraic equations, choose a suitable orientation for the wrench. Identify and label the cut points as before. Choose an origin and estimate the quantities you will need for your pseudocode. Finally, derive the equations and fill in the chart as in the previous task.

Cut	Pseudocode	Equation(s)

Once again, check your computations by graphing these equations on a graphing calculator.

PROJECT ACTIVITY:

Wrenches and many other objects come in various sizes. Now that you have designed one wrench, you can design another with a larger opening by adapting this design. For example, if you designed a wrench with a 2 cm opening, you can design one with a 3 cm opening by increasing the size of the entire design by 50%. To do this, first decide if it is total area you want to increase by 50% or total perimeter by 50% or some other characteristic of the wrench.

APPENDIX

In order to graph functions with domain restrictions in the TI-family of calculators, you can use the syntax

y=(function)/(restriction).

For instance, the equations for the three quarters disk would be:

y=(-x+10)/(x≥5 *and* x≤ 7)
y=x/(x≥5 *and* x≤ 7)
y= (5+ √(8-(x-5)^2))/(x≥2.17 *and* x≤ 7)
y= (5- √(8-(x-5)^2))/(x≥2.17 *and* x≤ 7)

Note: This is not precisely the way the calculator manual indicates to do domain restrictions, but it is more effective.

MATHEMATICS LABORATORY INVESTIGATION

NOISE POLLUTION

Topics: **Logarithmic functions and their graphs.**
Prerequisites: *Algebra II*
Equipment: Sound Meter

I. SITUATION

The *Tube Steak* Restaurant has recently moved into the neighborhood. This restaurant has a fan on the side of it which pulls the smoky, greasy air from inside the restaurant and blows it outside into the atmosphere. This fan operates in such a way that it emits a sound at a sound level of about 90 dB. This is about the same sound level as a large truck so, not unreasonably, the neighbors find the sound annoying at best. Suppose that the restaurant finds that it needs a second fan, of equal sound level, to properly ventilate the restaurant. Naturally, the neighbors will be upset about this as a sound level of about 130 dB is painful to the human ear.

The problem facing the engineer is to do a calculation to determine what the total sound level of the two fans will be and to convince the neighbors that the second fan will not be a major problem for them.

II. BRIEF INTRODUCTION TO THE SCIENCE CONCEPTS:
(See the Appendix for more details)

Sound is transmitted through air (or any other medium) by the movement of the molecules of that medium. As a tuning fork vibrates, for example, it causes the alternating compression and rarefaction of the air molecules surrounding it. These compressions and rarefactions move outward at a constant rate (*the speed of sound* in that medium) and have a spherical shape, much like the waves that are produced when a pebble is dropped in a puddle of water, the difference being that these "waves" travel outward in three dimensions rather than two.

The compression of the air molecules causes an increase in the density and pressure of the air. Likewise, the rarefaction causes a decrease in the density and air pressure. It is these changes in air pressure (measured as the *root mean squared pressure*, p_{rms}, in units of Pascals, Pa) that our ears detect and which cause us to hear the sound.

Travelling sound waves transmit energy. The rate at which this energy is tranmitted is called the *power, W*, (measured in Watts, W) and the amount of power measured over a unit of area is called the *intensity, I*, (measured in W/m^2). The intensity is proportional to the square of the pressure, p_{rms}.

Pressure has a tremendous range of values, from the threshold of hearing, 0.00002 Pa, to that produced by a rocket launch, 200 Pa. In order to measure values over this range, a logarithmic scale is used. The *Sound Intensity Level, L_I*, is defined to be the logarithm (base 10) of the ratio of the sound intensity to the intensity of the threshold of hearing,

$$L_I = \log\frac{I}{I_0} \qquad \text{(Eq. 1)}$$

where I_0 is the least sound intensity detectable by the human ear (10^{-12} W/m^2). The units of sound intensity level are called bels. In order to get a better scale, bels are broken down into tenths, called decibels (dB). The formula for sound intensity level in decibels is therefore

$$L_I = 10\log\frac{I}{I_0} \qquad \text{(Eq. 2)}$$

Since the intensity is proportional to the square of the pressure, equation 2 can be written as

$$L_I = 10\log\frac{p^2}{p_0^2} \qquad \text{(Eq. 3)}$$

or

$$L_I = 20\log\frac{p}{p_0} \qquad \text{(Eq. 4)}$$

where p_0 is the sound pressure at the threshold of hearing (20 microPascals or 20μPa). The quantity on the right hand side of Equation 4 can be defined as the *Sound Pressure Level.* With this definition, the sound pressure level and the sound intensity level will always give the same number of decibels, so, in fact, it doesn't matter whether our sound meter measures sound pressure or intensity levels.

From this point on, we will refer only to the *Sound Level, L,* understanding that it is the common value of L_I and L_P. For instance, equation 2 becomes

$$L = 10\log\frac{I}{I_0} \qquad \text{(Eq. 2')}$$

III. SOUND LEVELS

Task 1: Measuring Sound Levels

Sound Meter Instructions:

1. When comparing the sound levels of objects measured using sound meters, be sure to keep the distances from the objects to the meters constant -- about four feet will do.
2. If you have a digital sound meter, the reading will keep changing (due to background noise, etc.). Take a reading every second for about 10 seconds and average your results. Give your result to the nearest tenth of a decibel.
3. If you have an analog sound meter, read it to the nearest decibel.

Take two similar objects which emit a fairly quiet noise (two of the same model of fan, for example). Using a sound meter, measure the sound level of each fan individually and both fans together.

How does the sound level of both fans running simultaneously compare with the sound levels of the individual fans?

	Digital Meter Readings			Analog Meter Readings	
	1 object	**2 objects**		**1 object**	**2 objects**
Sound Level					

Repeat this measurement with two of the same type of objects which emit a fairly loud noise (bells, vacuum cleaners, or hair dryers, for example).

	Digital Meter Readings			Analog Meter Readings	
	1 object	**2 objects**		**1 object**	**2 objects**
Sound Level					

(While you are taking measurements, you may wish to skip to Task 3 before doing Task 2.)

Task 2: The Algebra of Sound Levels

The object of this task is to determine algebraically what happens to the sound level when the sound intensity is doubled.

Expand the right hand side of equation 2' as a sum of simpler logs using properties of logarithms.

Replace the intensity, *I*, by *2I* in Equation 2' and expand using properties of logarithms.

Numerically, what is the difference between these two expressions?

How does this compare with what you observed in Task 1?

Task 3: Combining Unequal Sound Levels

Take one object which emits a fairly quiet sound and one which emits a fairly loud sound. Using your sound meter, examine the decibel level of the individual objects versus that of the objects running simultaneously.

	Digital Meter Readings				Analog Meter Readings		
	loud object	**quiet object**	**both objects**		**loud object**	**quiet object**	**both objects**
Sound Level							

Task 4: The Algebra of Unequal Sound Levels

Use the sound level of the quiet object measured at the beginning of Task 1 (the digital reading if available) and equation 2' to compute the sound intensity, I, of the object. (Use 10^{-12} W/m^2 as I_0.)

Do the same for the loud object measured in Task 1. (Be sure to use the reading from the same sound meter as you used in the first part of this task.)

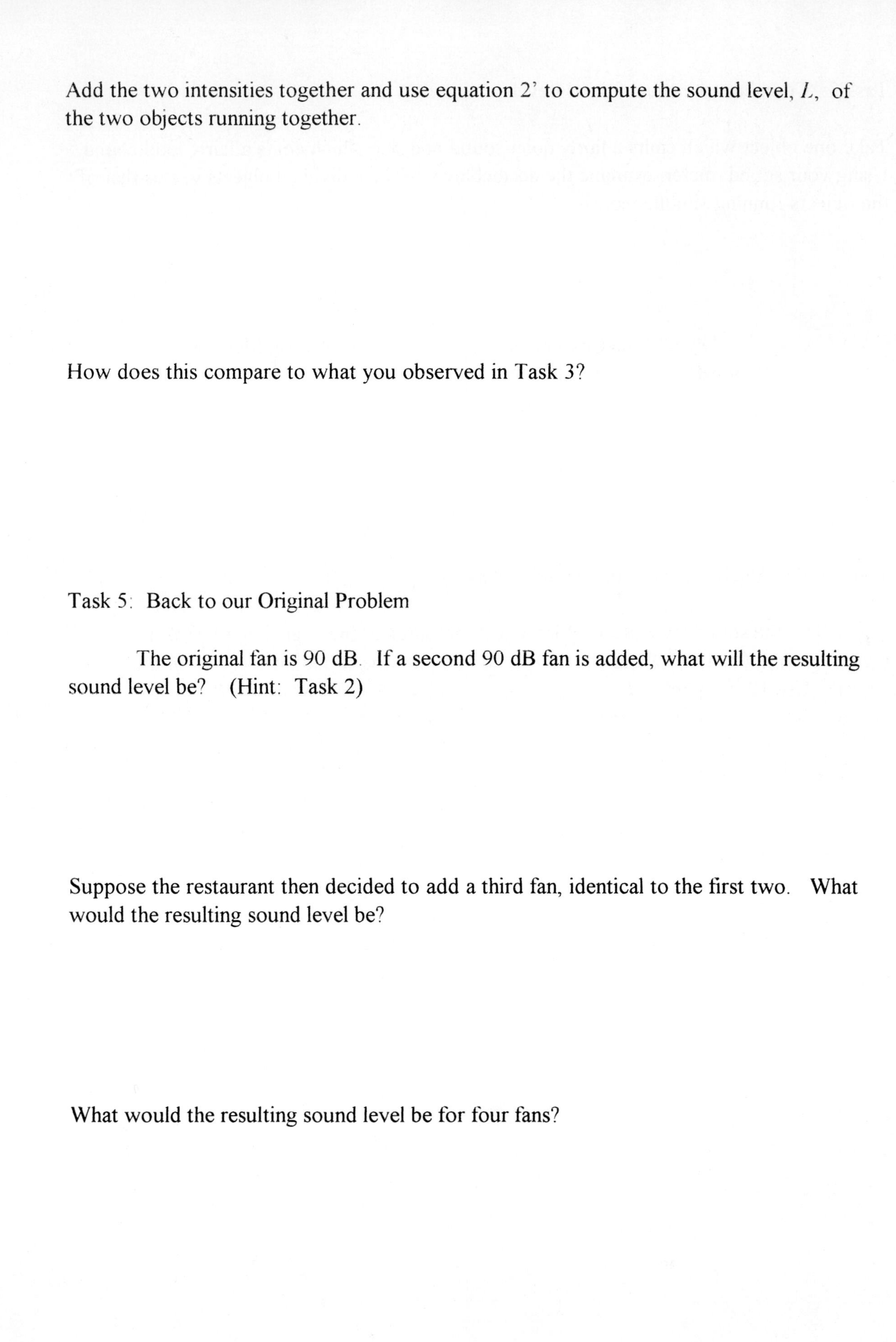

Add the two intensities together and use equation 2' to compute the sound level, L, of the two objects running together.

How does this compare to what you observed in Task 3?

Task 5: Back to our Original Problem

The original fan is 90 dB. If a second 90 dB fan is added, what will the resulting sound level be? (Hint: Task 2)

Suppose the restaurant then decided to add a third fan, identical to the first two. What would the resulting sound level be?

What would the resulting sound level be for four fans?

What would the sound level be for five fans? 10 fans?

Make a graph of thc number of fans versus the sound level. What characteristics does this graph have?

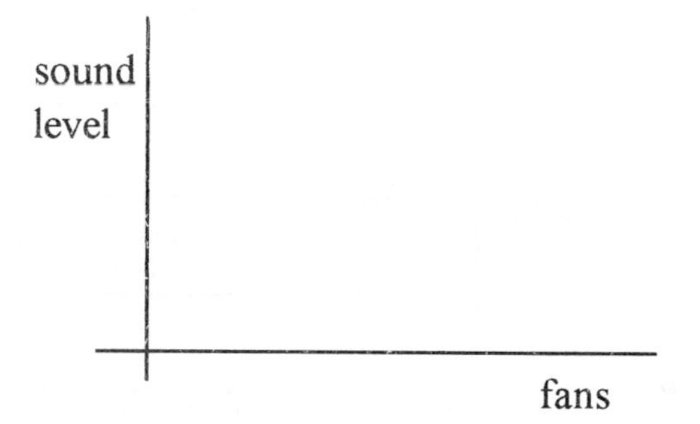

Suppose the restaurant wanted to know how many fans could be added before the sound level reaches up to the threshold of pain (130 dB)? (This, in fact, could never be done as other OSHA standards limit the amount of continuous noise exposure, but work it out anyway.)

If k denotes the number of 90dB fans that could be added, equation 2' tells us

$90 = 10\log\frac{I}{I_0}$ and $130 = 10\log\frac{kI}{I_0}$ where I is the intensity of the fan and I_0 is the intensity of the threshold of hearing, as before. Using properties of logs, solve this for k.

Task 6: A Rule of Thumb for Adding Sound Levels

The rule of thumb is a graphical one. To add sound levels M and N (where N is the smaller), look up M-N on the x-axis of this graph and read the quantity Y off the vertical axis. The sound level of M and N running simultaneously is then $M+Y$. (The graph is an exponential decay with y-intercept 3, decaying to $y=2$ by $x=2$ and to $y=1$ by $x=6$. By $x=20$, y is negligible.)

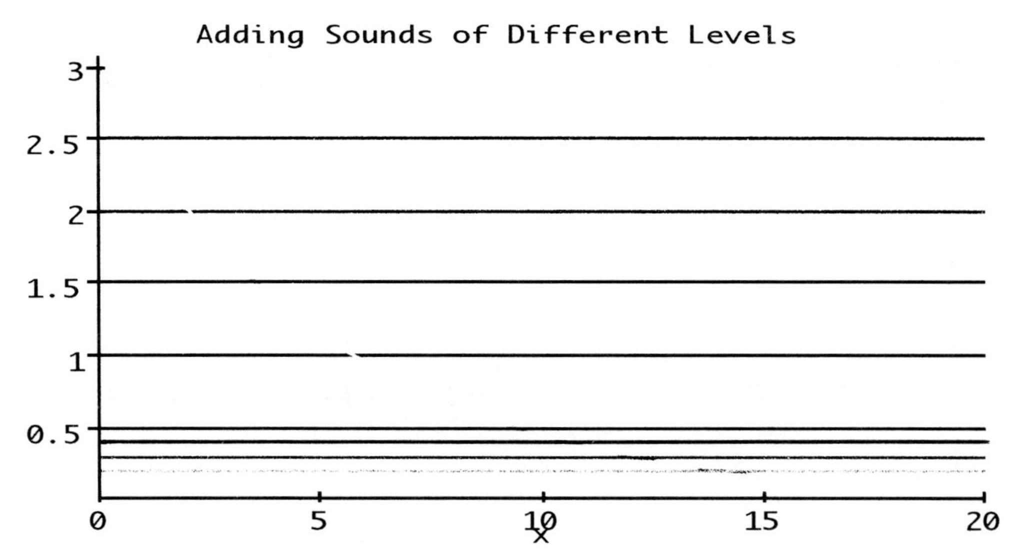

Example: 75dB + 80 dB would be about 81.2 dB. When x is 5 (80-75), y is about 1.2, so the sound level of both together would be 80 + 1.2 = 81.2 dB.)

Redo Task 1 using this rule of thumb. That is, fill in the following table using your measurements from Task 1 (digital meter if available) in the columns labelled **1 object** and use this rule of thumb to compute the entries in the columns labelled **2 objects**.

	quiet objects			loud objects	
	1 object	**2 objects**		**1 object**	**2 objects**
Sound Level					

Redo Task 3 using this rule of thumb. Again, use your measurements from Task 3 (digital, if available) to fill in the first two columns of this table and use the rule of thumb to compute the final entry.

	loud object	**quiet object**	**both objects**
Sound Level			

How does the rule of thumb compare compare to the results that you previously observed?

Suppose you have three objects running simultaneously (say 70, 80, and 90 dB respectively). Use the rule of thumb to add the first two of them and then to add the sum to the third object.

Repeat this by adding them in the order 70, 90, 80 dB. Does it make a difference which order you use?

(optional): Add the sound levels of the objects in the previous question algebraically (by calculating the intensities of the objects, adding the intensities, and then calculating the sound level of the result). How does this compare with the answer(s) you got by using the rule of thumb in different orders?

Task 7: The Shortcomings of the Rule of Thumb

Add sounds of 30 and 32 dB using the rule of thumb and then algebraically (as in Task 4). How do your answers compare?

Repeat with sounds of 100 and 102 dB.

Do the actual sound levels matter, or only the difference between them?

APPENDIX ON SCIENCE CONCEPTS:

As we remain still and sound "waves" pass by us, we notice the periodic rising and falling of the pressure in a regular fashion. This behavior is called *sinusoidal,* as it is easily modelled by a sine wave such as the following.

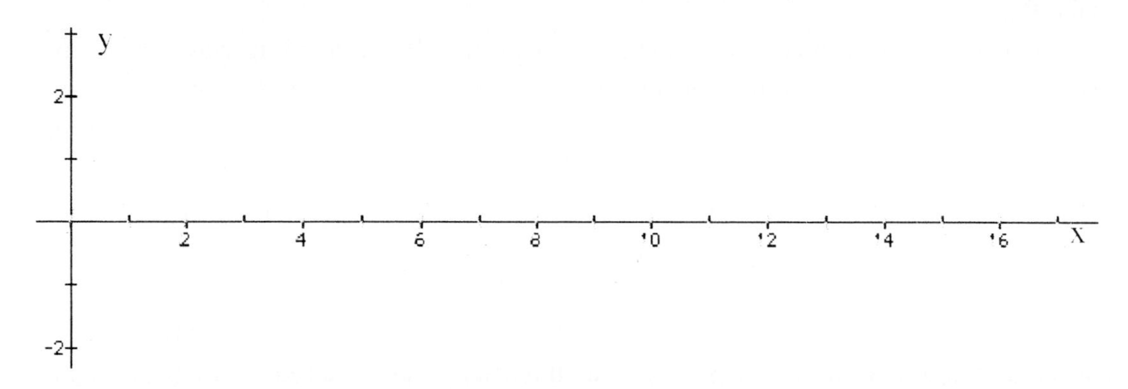

The *period, T,* of the wave is the time between consecutive occurrences of highest (or lowest) pressure. In the figure above, the period is 2 units on the x-axis, assuming that the horizontal axis is measuring time. The *wavelength* , λ, of a sound is the distance travelled by the sound during one period. The period and wavelength are related by the formula

$$\lambda = cT \quad \text{(distance = rate} \times \text{time)}$$

where c is the speed of sound in the medium. The *frequency*, f, of the sound is defined to be the number of pressure "peaks" passing by a fixed point in one second and so is the reciprocal of the period.

$$f = \frac{1}{T} = \frac{c}{\lambda}$$

The *amplitude, A,* is the difference between the maximum and the average pressure (the amplitude is 1 unit in the figure above). It is the amplitude and the frequency that our ears sense in order to hear the sound.

MEASURING SOUND:

There are several ways of measuring sound. One is to try to quantify the pressure in a sound wave. Averaging the pressure will not do since, as the above example shows, the average pressure is not only zero, but is independent of the amplitude and the period, the two quantities our ears can sense. What is necessary is a mathematical way of computing the average pressure which does not allow this information to be lost. The method used is to compute the *root mean squared pressure,* p_{rms} . First, label the vertical axis so that the average pressure corresponds to zero. (The vertical axis is, in effect, measuring the difference in pressure from the average). This pressure differential function is then squared to prevent points of positive difference from "cancelling out" points of negative difference. Then, average this (squared) pressure differential function over one period, and take the square root of the result. This definition guarantees that root mean squared pressure includes the amplitude and period information necessary to analyze the sound. In the metric system of measurement, pressure is measured in Pascals, Pa.

A second way to measure sound is by its *power, W*. Travelling waves of sound transmit energy in the direction of propagation of the wave. (referred to in physics as *work*.). The rate at which this work is done is defined to be the power and is measured in Watts, W.

A third way to measure the sound is by the *intensity, I*. The intensity is the average sound power per unit area perpendicular to the direction of propagation of the wave and has units W/m^2.

These three measures of sound are all interrelated. Intensity is merely power measured over a unit area ($I = W/$ area). The intensity is also related to the root mean squared pressure by the formula

$$I = \frac{(p_{rms})^2}{\rho c} \qquad \text{(Eq. A1)}$$

where ρ is the density of the medium (in kg/m^3) and c is the speed of sound in the medium (in m/sec).

LEVELS AND DECIBELS

A problem with all of these measures is that they have values over a tremendous range. The human ear can sense pressure of as little as 0.00002 Pascal while a rocket launch is greater than 200 Pa. In order to work with quantities of such huge variation, it is helpful to use a logarithmic scale. We define the *Sound Intensity Level*, L_I', to be the logarithm (base 10) of the ratio of the sound intensity to a reference intensity. That is,

$$L_I' = \log \frac{I}{I_0} \qquad \text{(Eq. A2)}$$

where I_0 is the reference intensity, the threshold of hearing. The units of sound intensity level are bels, named after Alexander Graham Bell. In order to distinguish these levels more clearly, bels are broken into tenths, called decibels, and denoted dB. The formula for sound intensity level in decibels is given by

$$L_I = 10 \log \frac{I}{I_0} \qquad \text{(Eq. A3)}$$

The units are sometimes denoted dB (re: $10^{-12} W/m^2$) to indicate the value of the reference intensity. Similarly, we can define sound power level by means of a similar equation.

$$L_W = 10 \log \frac{W}{W_0} \qquad \text{(Eq. A4)}$$

where the base sound power is chosen as that of the threshold of hearing (10^{-12}W). Since sound power and sound intensity are proportional and since, in both cases, we have chosen the threshold of hearing as our reference, a sound having a power level of x dB (re. 10^{-12}W) will also have an intensity level of x dB (re: 10^{-12}W/m^2).

This correlation can be carried a step further. Since, by Equation 1, the sound intensity is proportional to the square of the sound pressure, Equation 3 becomes

$$L_I = 10\log\frac{p}{p_0^2} \qquad \text{(Eq. A5)}$$

where p is the root mean square pressure and p_0 is a reference pressure. This equation then becomes

$$L_I = 20\log\frac{p}{p_0} \qquad \text{(Eq. A6)}$$

If we choose p_0 to be the sound pressure at the threshold of hearing (20 microPascals or 20μPa), we can define the quantity on the right hand side of equation 6 to be the *Sound Pressure Level, L_P*. Using this definition, the sound pressure level, the sound intensity level, and the sound power level will all give the same number (re: the threshold of hearing). In this way, referring to a sound as being 80 dB is unambiguous, even if we don't indicate to which of the three sound levels we are referring.

MATHEMATICS LABORATORY INVESTIGATION

QUALITY ASSURANCE IN MACHINING

Topic: **MEAN, STANDARD DEVIATION, TOLERANCE, SUMMATION NOTATION**

Prerequisite knowledge: *Basic Algebra, Scatterplots*

I. INTRODUCTION:

The products of many types of machines, including milling machines and lathes,must be checked periodically to determine whether the machine is producing items that are the correct size. If the produced part is not within proper tolerances, the process is said to be out of control and must be adjusted.

II. WHY DO TOLERANCES HAVE TO BE SO PRECISE?

Many parts, particularly metal ones, expand very little when heated. Expansion is sometimes imperceptible to the human eye. Then, equipment such as a micrometer or calipers may be required to discern the expansion.

When parts are put together, the fit cannot be too tight or too loose. If too loose, one part may fall out. If too tight, the piece may not operate correctly. Welding parts together is not always an option for many reasons: parts may have to fit inside one another, which makes it difficult (if not impossible) to weld from the inside; or the parts themselves may not be able to withstand the extreme heat generated by welding.

Tolerances are generally small to obtain a precise fit but must be large enough to allow for heat expansion (and/or contraction in cold conditions.)

III. READING PART SPECIFICATIONS FROM THE DRAWING

On the drawing shown here are client specifications. One of those specifications is the **tolerance**. Tolerance values show the permissible deviation from a specified value of a structural dimension. One way is to have the part dimension listed with a ± after it as in the drawing to the right. This drawing has tolerances measured in thousandths of the measurement units. The closer the tolerance, the more costly the process because of the degree of accuracy involved. The manufacturer must meet the client's specifications to within the tolerance range. The specifications and tolerance assure the part will fit precisely into a larger design. The most common way manufacturers can assure this is by setting the machine limits within the boundaries of the client's tolerances.

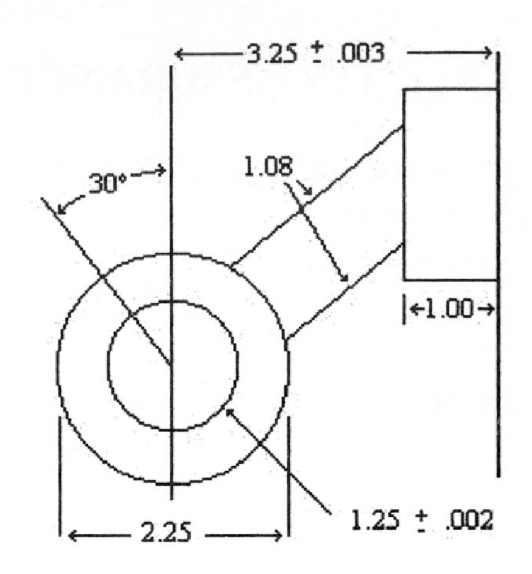

What is the client's minimum acceptable diameter for the inner circle? The maximum? What is the client's acceptable range of dimensions for the horizontal distance from the center of the circle to the end of the piece?

IV. USING THE TOOLS OF THE TRADE

The tools of the trade are micrometers, calipers and mathematics!

You will be using either a micrometer or calipers, depending on the object you are measuring, to determine precise dimensions of the object. The units on these instruments are very finely calibrated.

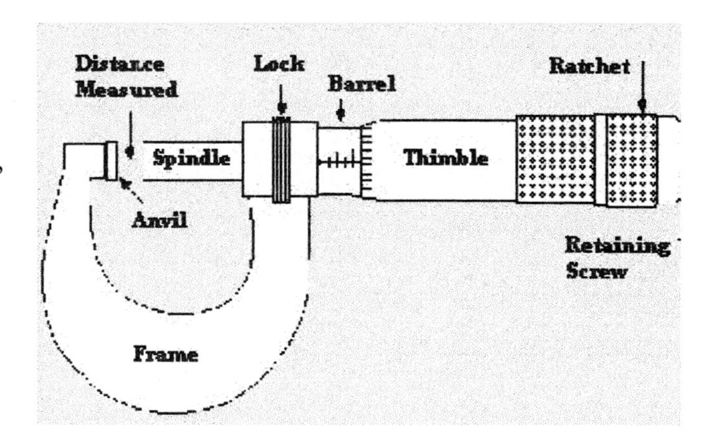

TASK 1: Calculating the mean diameter

Take at least 5 measurements of the diameter along the length of your object with a micrometer [or caliper.] Remember, the diameter goes through the center of a circular object. Devise some method to be reasonably certain your diameter measurement does pass over the center.
Describe your method here.

Record your results based on variation observed in the table below Be sure to indicate units.

measurement 1	measurement 2	measurement 3	measurement 4	measurement 5
x_1 =	x_2 =	x_3 =	x_4 =	x_5 =

Compute the average diameter. This is done by adding up all your measurements and dividing by the number of measurements you recorded. This value is called the ***mean*** of your measurements.

The formula for the mean is generally written in ***summation*** notation.

$$\bar{x} = \frac{1}{N}\sum_{i=1}^{N} x_i \quad = \_\_\_\_\_\_\_\_\_\_$$

The upper case Greek letter *sigma*, Σ, is used for summation. The *i* is an index to count the number of values added up. The *N* is the number of values to add. The bar over the x indicates it is the average or mean value.

The calculation of the mean you just performed was $\bar{x} = \frac{x_1 + x_2 + x_3 + x_4 + x_5}{5}$.

Write this calculation using summation notation.

It is often useful to look at data graphically. This can give a good sense of how your measurements varied.

TASK 2: Representing the data graphically

Plot the mean value you calculated at the center of a number line. Then, plot your measured data points on the same number line using a constant scale. Another way to say this is be sure to choose division points so all the measured data will fit on the line and are not crowded together.

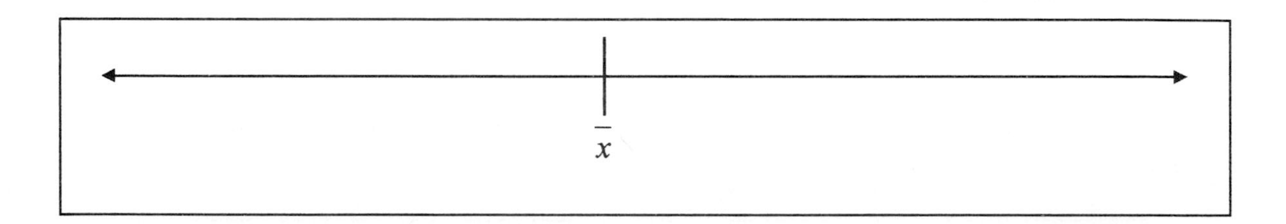

Look at the object you used to take the diameter measurements. Think about what the data points mean in terms of the shape of the object. What does the deviation in diameter measurements tell you about the shape of the object? Write a few sentences to explain your understanding of the situation.

You should now have some understanding of how important the deviation from the mean is. It tells you about the shape of the object you took measurements from. You will next come up with some comparative number that tells you something about how far your measured data deviates from the mean diameter.

A measure of spread of the data around the mean value is called ***standard deviation***. The formula for the standard deviation of measured samples looks very complicated but it is just a sort of averaged distance of all the data points from the mean value.

TASK 3: Calculating the standard deviation of your measured data

If x_i is one of your data points, its distance from the mean value is $x_i - \bar{x}$

Calculate the distance of each of your data points from the mean diameter value by subtracting the mean value from it. Enter the distance values below. Some distances will be negative if the measured data point is smaller than the mean.

	$(x_1-\bar{x})$	$(x_2-\bar{x})$	$(x_3-\bar{x})$	$(x_4-\bar{x})$	$(x_5-\bar{x})$
Deviation (x_i-x)					

Since we will want to account for all the deviations from the mean, we do not want negative distances to cancel out positive distances. So, all the distances are squared to yield positive numbers before adding them.

Square the above distances and enter them in the next table.

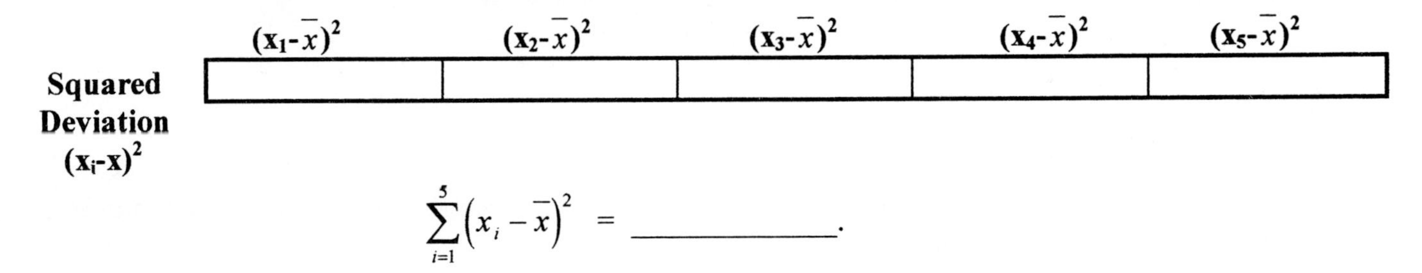

	$(x_1-\bar{x})^2$	$(x_2-\bar{x})^2$	$(x_3-\bar{x})^2$	$(x_4-\bar{x})^2$	$(x_5-\bar{x})^2$
Squared Deviation $(x_i-x)^2$					

$$\sum_{i=1}^{5}\left(x_i - \bar{x}\right)^2 = \_\_\_\_\_\_\_\_\_\_.$$

This last result is the sum of the squares of the distances of the sample data from the mean.

To find the mean of these "squared deviations," you would normally divide by N. However, dividing by (N-1) instead will produce a larger value that will be a a better estimate of the true variation of <u>all</u> similar machined parts. (It can be shown that division by N-1 gives a better "true average" of the squared deviations.) This "mean-squared-deviation" is called the *variance* of the data.

$$\textbf{Variance=} \quad \frac{\sum_{i=1}^{N}\left(x_i - \bar{x}\right)^2}{N-1} = \_\_\_\_\_\_\_\_\_\_.$$

While the variance is useful for some statistical purposes, it is difficult to apply directly as a measure of variability, because now the units are squared. So, if you take the square root of the variance you get a number that again measures a distance.

This distance is called the ***standard deviation***, and is denoted by the letter S. This is sometimes represented as σ_{n-1} on some computers or calculators (using the lower case Greek letter *sigma*).

$$S = \sqrt{\frac{\sum_{i=1}^{N}\left(x_i - \bar{x}\right)^2}{N-1}}$$

Sometimes this number is referred to as the *root-mean-squared deviation* (or RMS deviation).

Calculate the standard deviation for your data.

$S =$ ________________.

Remember, this is a "kind of average" deviation of the data points from the mean. Another valuable feature of standard deviation is it makes it possible to compare different numbers of readings. If your group took five measurements and another group took six or any different number of measurements looking at each group's standard deviation allows comparison of variability independent of the number of data points collected.

V. DOES YOUR PART PASS INSPECTION?

Remember, you were measuring this part to determine if it met specifications. The micrometer or calipers were the tools that let you measure the object. The mean and standard deviation are the mathematical tools that will let you accept or reject the part.

TASK 4: PLOTTING YOUR RESULTS

On the number line, replot the mean value you calculated at the center of a number line. Replot your measured data points on the same number line. Enter in division points for values representing the following distances from the mean: $-3S$, $-2S$, $-1S$, $+1S$, $+2S$ and $+3S$. A typical tolerance for parts is +/- .002 as in the drawing in section III. Mark this tolerance on the same number line below that you marked with the standard deviation values. Again, think about fitting all the measured data on the line so that it is not crowded together, and remember to use appropriate units.

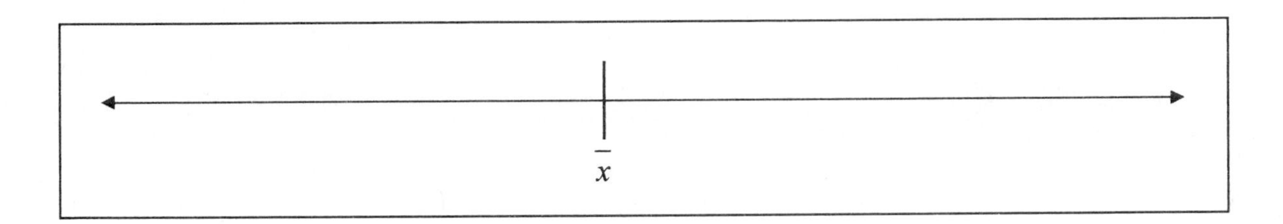

Did any of your measurements fall outside the tolerance requirements?

Most of your observations should have fallen within $\pm$ 3S, which is usually referred to as 3σ (even though it's only based on a sample of data). Manufacturers strive for a "3 Sigma" tolerance level to maintain process control and insure the machine is producing parts of acceptable quality.

VI. ADDITIONAL ACTIVITY

Exploration: With larger data sets the most reasonable way to do the calculations is by using technology. Most calculators, and many software programs, can compute standard deviations automatically from listed data. See if you can determine how to do this with your calculator, and then compare your results with those calculated above.

MATHEMATICS LABORATORY INVESTIGATION

ROBOTICS

Topic: **PARAMETRIC EQUATIONS**
Prerequisite knowledge: *Cartesian Coordinate Systems, Algebraic Equations*

I. INTRODUCTION

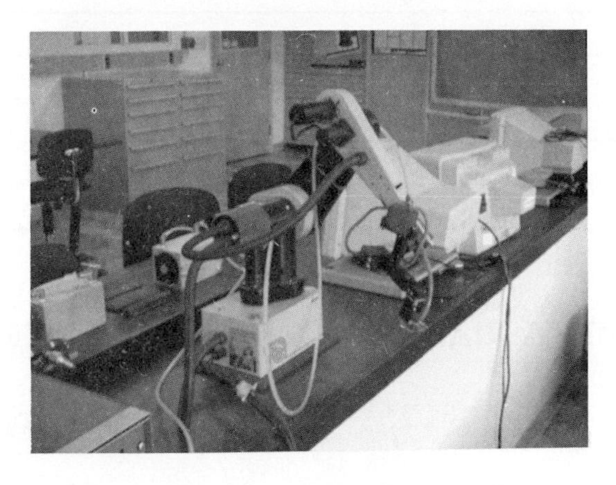

In an automated environment, there is often a need for more than one robot to be involved in a certain project. Because of the specialized nature of robots, each often being designed to perform a specific set of tasks, it is sometimes necessary for one robot to perform a task on an object and later for another robot to perform a different task on the same object. Somehow the first robot must leave the object where the other can retrieve it. For other projects, it may be necessary for the robots to pass parts to one another. If two robots are working independently on the same object, it is important that they do not collide. These synchronizations are accomplished mathematically by introducing variables, called *parameters*, in addition to the usual variables which describe the position of the robot. The resulting mathematical equations are called *parametric equations*. One example of a parameter would be a variable t introduced to represent time. The position equations could then be formulated as functions of t, giving instructions to the robot for not only where to go, but when to go there.

In another setting, many companies are now designing parts in parametric format for their machines. If design changes are necessary, all the designers have to do is change the value of the parameter and the other variables are adjusted automatically. In this format, for example, it is very easy to scale up or down the size of a part. This makes for a more economic use of labor and equipment.

II. CHARTING THE COURSE

TASK 1: Charting t values

The notation most commonly used for a two-dimensional position in terms of time is $x(t)$ and $y(t)$. For example, for a robot which etches a design onto glass (or a mirror), the position of the tool in the plane could be expressed by the equations $x(t) = t - 3$ and $y(t) = t$. Fill in the table below by finding values for $x(t)$ and $y(t)$ at the indicated values of t. (Even if t represents time, negative values are possible. For example, it could be measuring time in seconds before or after some event.)

t	$x(t)$	$y(t)$
-15		
-10		
-5		
0		
5		
10		
15		

III. VARIABLES WORKING TOGETHER

TASK 2: Plotting the three variables

To plot the points you computed in the previous table on the grid below, first choose a point to serve as the origin and then draw in the x and y axes. Then, plot and label the points (x, y) with their t values.

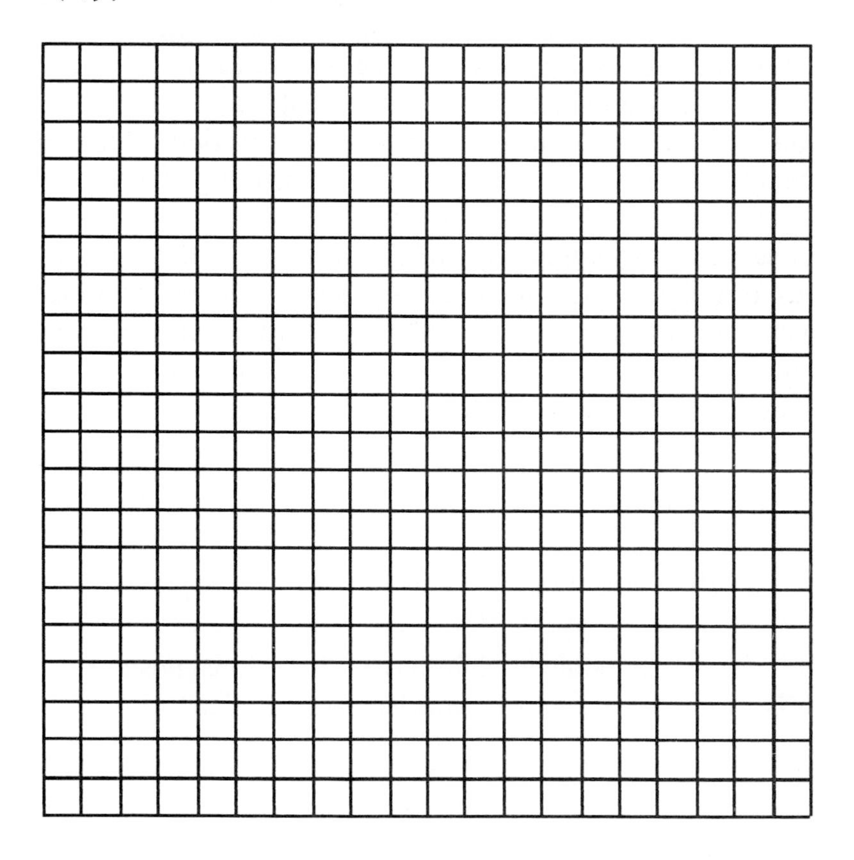

What, if any, pattern do you see in the points you just plotted?

TASK 3: Checking Your Work

Use a graphing utility to graph the equations. Make sure your graphing utility is in parametric mode. Check your window setting to make sure that all of the points from the table above appear. How does the graph correspond to what you did in Task 2? (If the graphs don't appear the same, try adjusting your calculator's window.)

TASK 4: Writing Your Own Equations

In the example above, it is not hard to see that the value of y is always 3 more than the value for x, indicating that the graph in Task 3 should look like the (nonparametric) function $y = x + 3$. There are other possible equations for $x(t)$ and $y(t)$ which have the same property, however. For instance, consider the equations $x(t) = 2t - 3$ and $y(t) = 2t$. Although the graph of these equations appears the same as the graph of the previous equations, examine the graphs at different values of t and explain the difference between the two sets of equations.

Suppose the path described by the equations in Task 1 needs to be shifted so that the etching tool of the robot goes through the location $(x, y) = (5, 9)$ along a line whose graph looks like $y = x + 4$. Write a set of parametric equations to accomplish this.

$x_1(t) =$

$y_1(t) =$

Check to see, by using a graphing utility, if the equations you made up do indeed go through the correct point.

At what time does your robot go through that location?

Is your solution unique? Explain why or why not.

TASK 5: Intersections

There are different kinds of glass etching robots. The simplest kind is a *Two-Dimensional Cartesian Manipulator* (as diagrammed below).

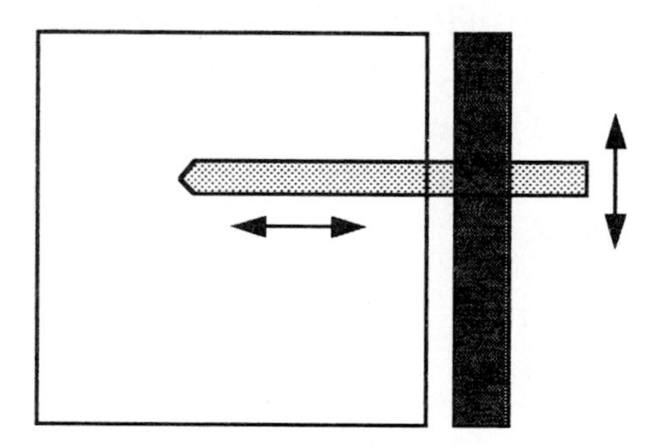

This is a diagram of the top view of the manipulator. The glass is represented by the white square. The manipulator itself consists of two major components (called *links*). The fixed link (shown here as the dark rectangle) is mounted on a table next to the glass. The movable link (shown here as the light shape) sits on top of the fixed link and slides (vertically in this picture) along the fixed link as well as across the fixed link (horizontally in this picture). The etching tool (obscured by the movable link in this view) hangs down from the end of movable link and etches the glass at a point directly underneath the pointed tip of the movable link.

It is not hard to see why this is called a Cartesian manipulator. If some point is chosen to be the origin, the amount by which the movable link slides across the fixed link determines the x coordinate of the tool. Similarly, the amount by which the movable link slides along the fixed link determines the y coordinate of the tool.

Suppose two robots are etching the same piece of glass simultaneously.

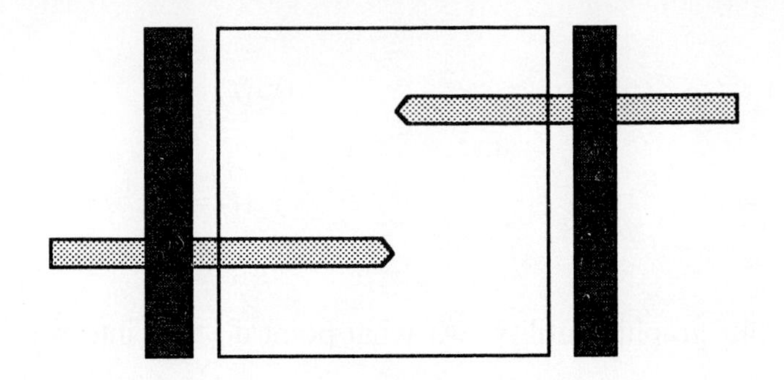

Suppose further that the curves that they are etching both go through the same location (5,9). Use your equations from Task 4 as the path of the lefthand robot. Create a second set of parametic equations $x_2(t)$ and $y_2(t)$ that also go through the point (5,9) and let them represent the path of the righthand robot. (Be sure that the second set of equations represents a different curve than the first set.)

$x_1(t) =$ $\qquad\qquad$ $x_2(t) =$

$y_1(t) =$ $\qquad\qquad$ $y_2(t) =$

Put the equations into your graphing utility. (If you are using a graphing calculator, you may need to make sure that it is in simultaneous mode.) At what values of t do they each go through the point (5,9)?

Do they go through the point at the same time or different times?

When they go through at the same time the crossing is called a *simultaneous* crossing. If they don't go through the point at the same time it is called *non-simultaneous*. An example of a simultaneous crossing is:

$x_1(t) = t$ $\qquad\qquad$ $x_2(t) = t$

and

$y_1(t) = t + 2$ $\qquad\qquad$ $y_2(t) = 2t$

Graph these using your graphing utility. At what point do they intersect? At what value of t? Describe how you found this.

An example of a non-simultaneous pair is given by the equations:

$$x_1(t) = t - 2 \qquad\qquad x_2(t) = t$$

and

$$y_1(t) = t \qquad\qquad y_2(t) = 2t$$

Graph these using your graphing utility. At what point do they intersect?

Which pair gets there first?

How much sooner (if t is measured in seconds)?

Why is the distinction between simultaneous and nonsimultaneous crossings important in this glass cutting application?

TASK 6: Further Considerations

Even if curves do not intersect, problems can result from two robots working at the same time. Consider the example in the following diagram. In this example, there are two robots etching simultaneously. The lefthand robot is etching a vertical line upwards on the glass. The right hand robot is etching a vertical line downwards on the glass.

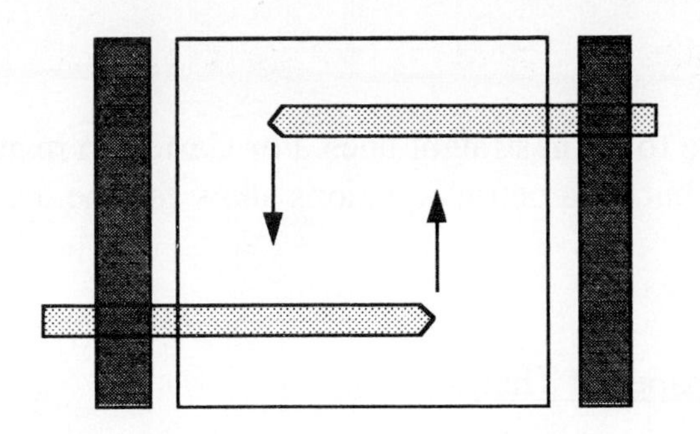

If the origin of the workspace is chosen to be the lower left corner of the glass and if the glass measures 24 inches by 24 inches, the equations for the two etched lines could be:

$$x_1(t) = 16 \qquad \text{and} \qquad x_2(t) = 8$$

$$y_1(t) = t \qquad\qquad\qquad y_2(t) = 24 - t$$

where the first pair of equations is for the robot on the left, the second pair is for the robot on the right, x and y are measured in inches and t is measured in seconds ($0 \le t \le 24$). What is the problem with this scenario?

Using only the equations, how would you detect this problem?

How would you solve this problem without changing the lines to be etched and still finish in 24 seconds? (Can you come up with more than one way?)

IV. CURVED PATHS

Robots don't have to go in straight lines. For Cartesian manipulators, straight lines are usually fairly simple, but parametric equations allow for the description of many different shapes.

TASK 6: Shapes of Things

Plot (either by hand or with a graphing utility) the curve $x(t) = t$, $y(t) = t^3$. What kind of shape is this?

Plot the curve given by $x(t) = 2\cos t$, $y(t) = 2\sin t$. (Make sure your grapher is in radian mode and the window is "squared".) Find the x and y intercepts of this curve. What kind of shape is this?

For the first curve, the equation for x can be substituted into the equation for y to come up with $y = x^3$. Although the same substitution technique won't work for the second curve, do you know a nonparametric equation for it?

Describe, both by a sentence and by a (nonparametric) equation, the curve given by $x(t) = t + 1$, $y(t) = t^2$.

Do the same for $x(t) = t^2$, $y(t) = 3\sin t$?

Using a graphing utility, examine the intersections of the following curves.

$$x_1(t) = (t - 1)^2 \qquad x_2(t) = t$$

$$y_1(t) = t \qquad y_2(t) = 3 \sin t$$

How many intersection points are there? What are they?

Are the intersections simultaneous or not? How did you determine this?

MATHEMATICS LABORATORY INVESTIGATION

SHEAR AND BENDING MOMENT I

Topic: **PIECEWISE-DEFINED FUNCTIONS**
Prerequisite knowledge: *Determining the equation of a line given 2 points*

I. INTRODUCTION:

The concepts of shear and bending moment are critical to the study of structures. To help you to understand these concepts picture a building frame that is subject to a very high wind.

The vertical columns will *bend* (a) and the floors will tend to slide or *shear* relative to the other floors (b). The actual deflection of the building includes both of these effects.

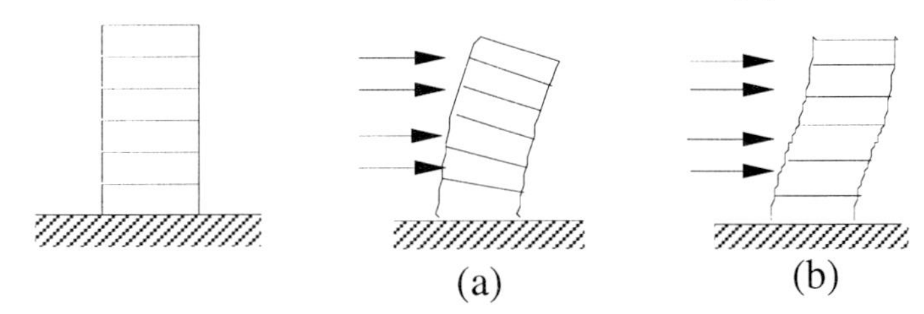

The *bending moment* at a certain point in a beam measures the tendency of the beam to bend at that point. The *shear* at a certain point in a beam measures the tendency of molecules in the beam to slide relative to the adjacent molecules. In order to investigate these two concepts, *reactions* must first be briefly considered.

When a beam is placed across two supports reactions occur that permit the supports to hold up the beam. This investigation involves a fairly simple structure, so finding thc rcactions will be straightforward.

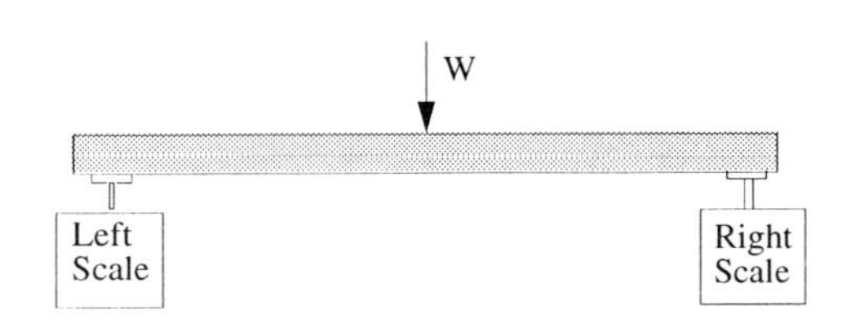

Using two scales to support the beam place a weight at the center of the beam. Read the reactions on the scales.

Weight: _______
Reaction at left scale: ______
Reaction at right scale: _____

Try this with several different weights placed at the center of the beam. In general if any weight is placed at the center of the beam, how can the reactions be calculated?

II. SHEAR

"Tension tends to move the particles of the material apart; compression pushes them together, shear makes them slide one with respect to the other." - Mario Salvadori in Why Buildings Stand Up

Shear is a vertical force (denoted V) which exists at every point in the beam. A shear diagram is a graph that shows $V(x)$, the shear at any point along the beam. Following mathematical convention, set x=0 at the left hand side of the beam and measure positive x to the right. In order to create a point on a shear diagram, choose a position and "cut" the beam there. Add up the vertical forces on the beam from the left endpoint to the cutoff point. Consider a 20 ft beam resting on two supports (this is called simply supported) with a 100 lb weight at the center.

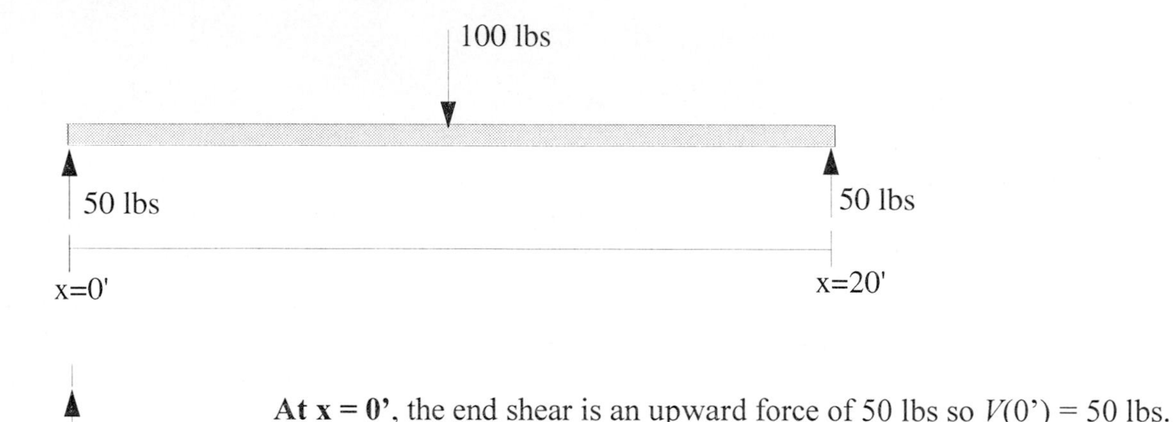

At x = 0', the end shear is an upward force of 50 lbs so V(0') = 50 lbs.

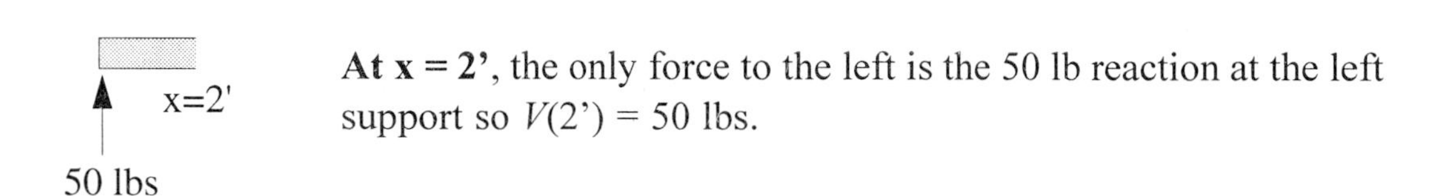

At x = 2', the only force to the left is the 50 lb reaction at the left support so V(2') = 50 lbs.

What is the shear at x = 5', x = 8', and x= 9.9'?

V(5') = __________ V(8') = __________ V(9.9') = __________

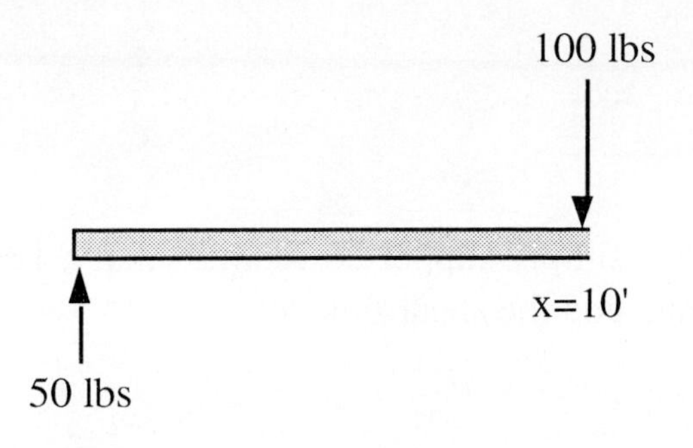

It isn't until x = 10' that another force is taken into account. Both the upward force of 50 lbs and the downward force of 100 lbs affect the shear from this point on. For x = 10':

V(10') = (+50 lbs) + (-100 lbs) = -50 lbs.

Find the shear at x = 10.2', 15', 18.75' and 20'.

V(10.2') = ________ V(15') = __________

V(18.75') = ________ V(20') = __________
(Be careful on this one.)

A plot of these points begins to form the *shear diagram* .

The actual shear diagram shows the shear at any point along the beam.

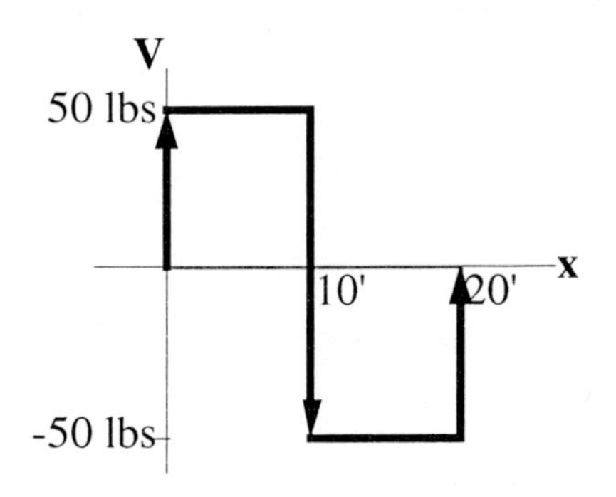

Its interpretation is that between 0' and 10' there is a constant upward shear of 50 lbs. Between 10' and 20' there is a constant downward force -50 lbs. At 0', 10', and 20' there are changes in shear. These are called points of *discontinuity*.

The mathematical function which represents the shear for this beam is

$$V(x) = \begin{cases} 50 \text{ lbs} & \text{if } 0' \leq x < 10' \\ -50 \text{ lbs} & \text{if } 10' \leq x < 20' \\ 0 \text{ lbs} & \text{if } x = 20' \end{cases}$$

It is an example of a *piecewise-defined* function. That is, a function that has different mathematical descriptions over different intervals in its domain.

III. THE SHEAR DIAGRAM

TASK 1.

A 180 lb person sits at the center of a 6 ft simply supported beam so that a 180 lb load is placed at $x = 3$ ft. Fill in the shear diagram and the shear function.

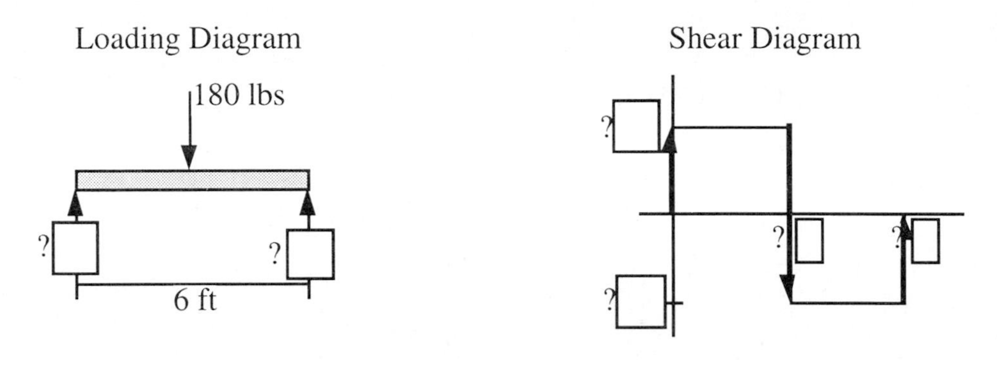

Shear Function

$$V(x) = \begin{cases} \_\_\_\_\_\_\_\_ \; if \; \_\_\_\_ \leq x < \_\_\_\_ \\ \_\_\_\_\_\_\_\_ \; if \; \_\_\_\_ \leq x < \_\_\_\_ \\ \_\_\_\_\_\_\_\_ \; if x = \_\_\_\_ \end{cases}$$

TASK 2:

What if the 180 lb load could be divided into two equal loads and placed equally spaced along the beam? Fill in the shear diagram and create the shear function.

Loading Diagram

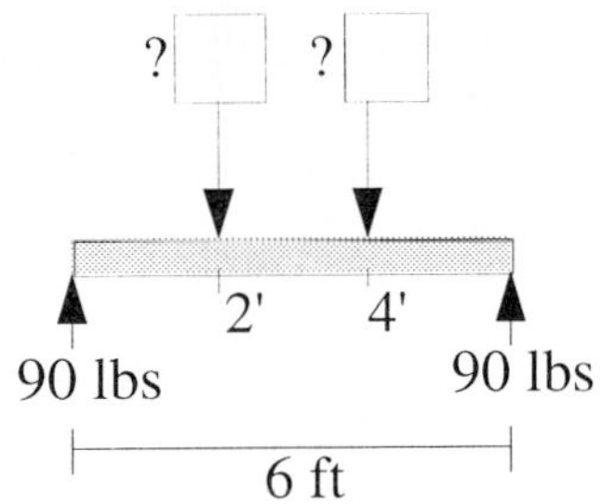

Shear Diagram

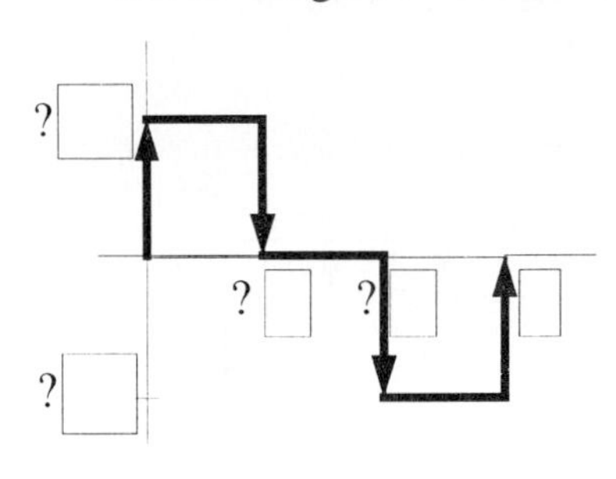

Shear Function

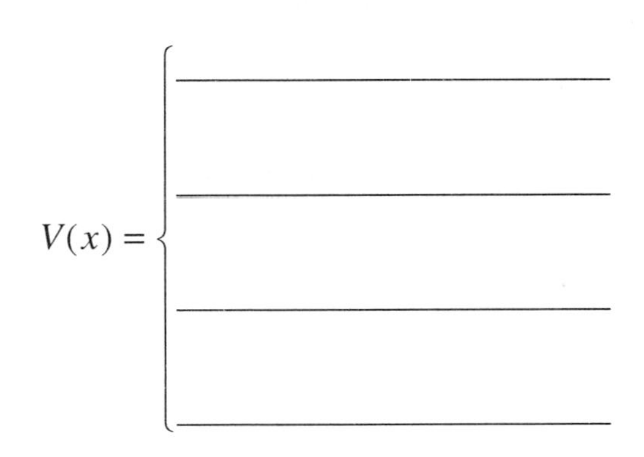

TASK 3:

Divide the 180 lb force into three equal loads, placed at equal intervals along the length of the beam. Draw the shear diagram and create the shear function.

Loading Diagram

Shear Diagram

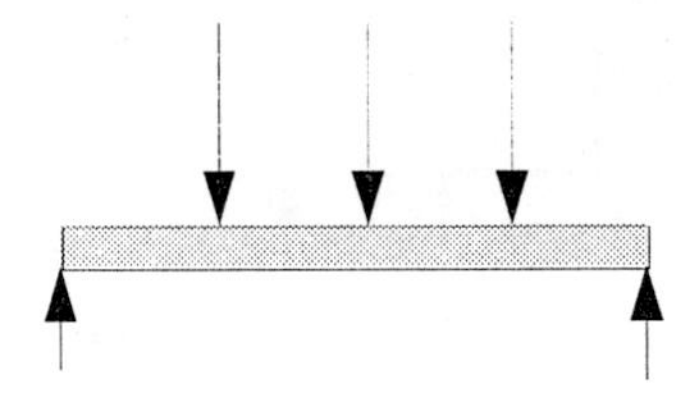

Shear Function

TASK 4:

Divide the 180 lb load equally at 1 ft intervals. What would the load be at each interval? __________

Draw the shear diagram and create the shear function for this situation.

Loading Diagram

Shear Diagram

Shear Function

TASK 5:

What if the 180 lb force were divided into 10 equal loads placed at equal intervals along the length of the beam? Draw a rough sketch of the shear diagram.

TASK 6:

If the 180 lb load were divided into 100 equal loads or 1000 equal loads equally spaced along the length of the beam, what would the shear diagram would look like? What would the analytical description of the shear function have to include?

What if we had the 180 lb load lie down on the beam so that the weight was *uniformly distributed*? How many equal loads has the 180 lb load been divided into?

What would the shear diagram look like?

Since the diagram is now composed of a smooth line (except for the isolated point (6', 0#)), we no longer need a piecewise-defined function to describe the shear diagram of this uniform load. Use your knowledge of linear functions to find the shear function for this loading situation.

$V(x) =$ ________________

By continually dividing a concentrated load into smaller loads and spreading them out we developed a function which can furnish information about the shear force of a uniformly distributed load on a simply supported beam. For example, what is the shear force at 3.75 ft from the left edge of the beam? ____________

Analyzing the shear diagram yields information about the location of important points along the beam. For example:

Where is the maximum shear? ___________________________________

Where is there no shear? ______________________________________

MATHEMATICS LABORATORY INVESTIGATION

SHEAR AND BENDING MOMENT II

Topic: **PIECEWISE-DEFINED FUNCTIONS**
Prerequisite knowledge: *Determining the equation of a line given 2 points, finding a quadratic equation using the regression feature on a calculator or computer.*

I. MOMENT:

The bending moment at a certain point in a beam measures the tendency of the beam to bend at that point. Wherever a beam is concave up (the ends of the section are curling up) we consider the moment to be positive. Where the beam is concave down we consider the moment to be negative.

positive moment   negative moment

Additionally, there are two elementary forces which can effect a beam: tension and compression. Tension occurs in an element when it is pulled on or stretched. Compression is a result of pushing or "compressing" an element.

Depending on where a material is used in a building it must be strong in tension and/or compression. Steel and wood are materials that are typically used in building construction because they are strong in both tension and compression.

TASK 1: Take a look at tension and compression.

Measure the length of a Styrofoam beam. Record the length here.

Draw vertical lines at equal intervals across one side of the beam.

Lay the beam on two supports and place a weight on the beam. It will deflect. The lines that you drew are no longer vertical, but they are still perpendicular to the top and bottom of the beam.

Measure the length of the top of the deflected beam and record it here: __________

Measure the length of the bottom of the deflected beam and record it here: _______

Is the top of the beam in tension or compression? Why?

Is the bottom of the beam in tension or compression? Why?

Is the bending moment at any point in the beam positive or negative? Why?

Somewhere in the center of the beam, its length did not change. This line is called the *neutral axis* or the *elastic curve*.

When an element bends, the amount of bending depends not only on the force that was applied but also on the material that the beam is made of, the shape it is cut into and the direction in which it is bending.

TASK 2: Not everything bends.

Instead of laying the beam across the two supports, loosely attach a string, chain, or elastic between two supports.

Apply a weight to the center of the string (or chain or elastic) and you will see that it doesn't bend. There is no point on the string where it is in compression or shortened. However, it has been stretched. The string is in tension only.

TASK 3:

Position the supports so that about 1/3 of the beam extends past one of the supports.

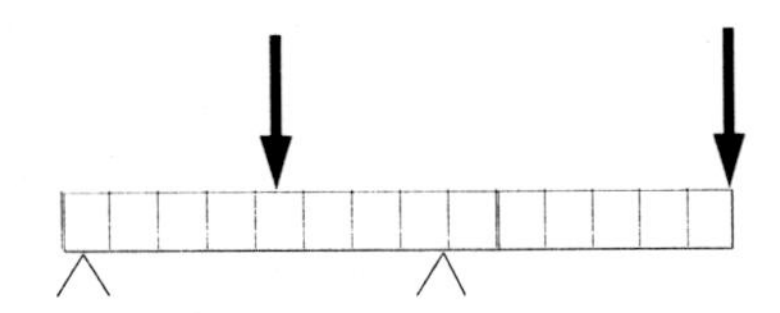

Place two weights on the beam: one between the two supports and one at the extended end of the beam. Sketch the elastic curve and indicate where the beam has positive bending moment and where the bending moment is negative.

Repeat the task using string (or chain or elastic) instead of the beam. What happened?

Is there any bending in the string?

This investigation will use string to illustrate some of the concepts relating to the bending that occurs in beams, but it is important to remember that beams and string are different. Weights applied to a beam result in both tension and compression while the same weights applied to a string cause only tension.

All along a beam we consider the tendency of the beam to bend.

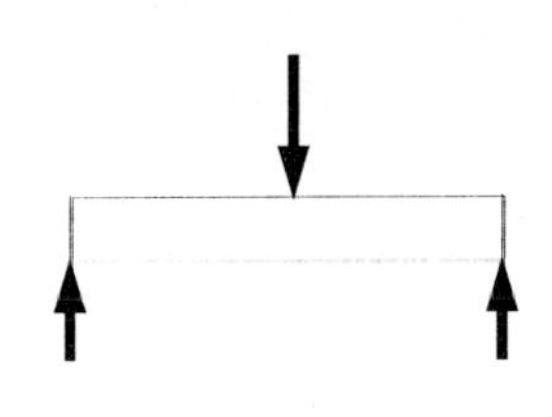

In the case of a simply supported beam it bends the most in the center and the least at the 2 ends. The way we quantify this tendency is to calculate the moment. The moment of a single load at a point is the product of the load times its distance from the point.

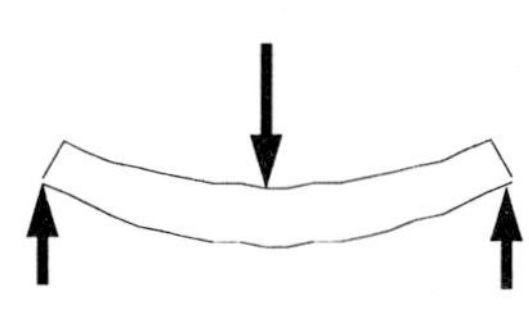

The moment at a point in the beam is the sum of the moments of all the forces to the left of that point. So if we consider the situation of a 100 lb load at the center of a 20' beam, we can calculate:

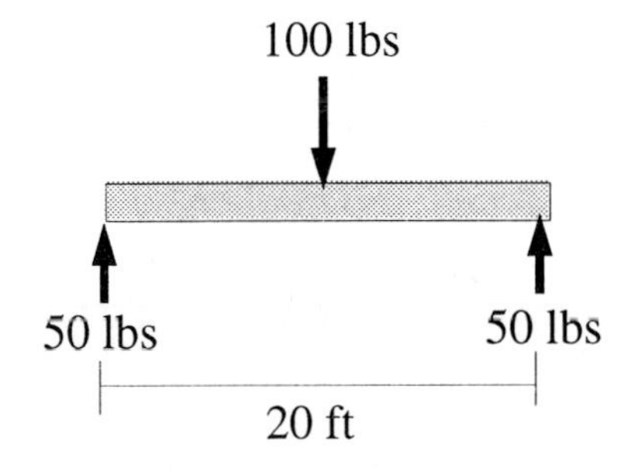

M(0') = (50 lbs)(0') = 0 ft lbs
M(8') = (50 lbs)(8') = 400 ft lbs
M(10') = (50 lbs)(10') + (-100 lbs)(0') = 500 ft lbs
M(13') = (50 lbs)(13') + (-100 lbs)(3') = 350 ft lbs
M(20') = (50 lbs)(20') + (-100 lbs)(10') + (50 lbs)(0') = 0 ft lbs

Use the examples above to help you to find the missing moments:

M(0') = 0 ft lbs

M(2') = _______

M(5') = ______

M(8') = 400 ft lbs

M(9.5') = _______

M(10') = 500 ft lbs

M(13') = 350 ft lbs

M(15') = _______

M(18') = _______

M(20') = 0 ft lbs

Plot these values and you will see a moment diagram taking shape.

M(x) in ft lbs

This diagram is composed of two separate line segments so its function will be piecewise-defined in two segments.

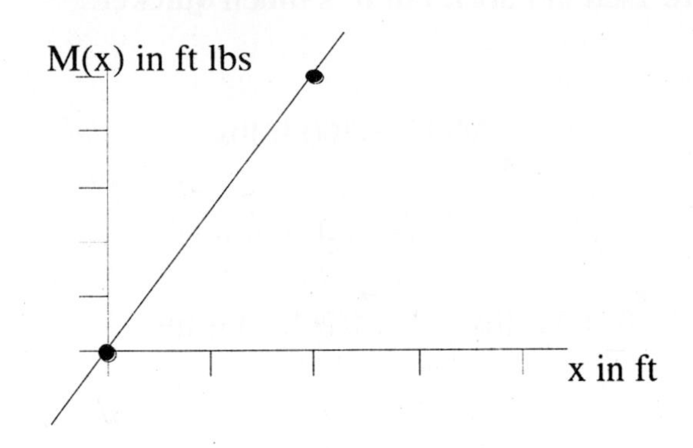

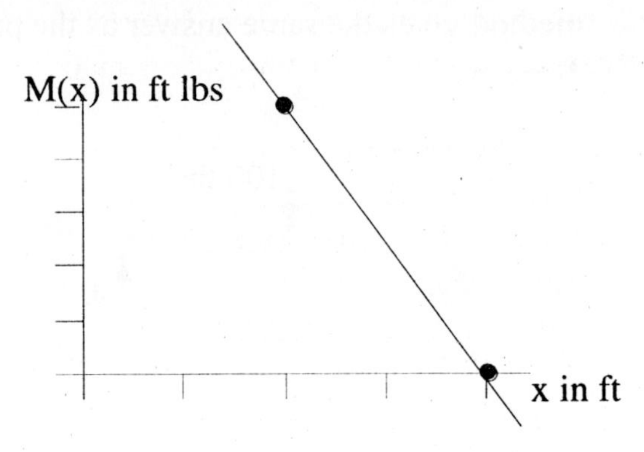

Find the equation of the line containing the points (0', 0 lbs) and (10', 500 lbs).

Find the equation of the line containing the points (10', 500 lbs) and (20', 0 lbs).

The function for the moment is

$$M(x) = \begin{cases} \_\_\_\_\_\_\_\_\_\_\_\_\_\_\_\_ \\ \_\_\_\_\_\_\_\_\_\_\_\_\_\_\_\_ \end{cases}$$

II. MOMENT DIAGRAMS

A model that assists in visualizing the moment diagram is created by hanging a chain, string or elastic band loosely between 2 supports. A weight hung at the center of the string pulls it taut and forms 2 line segments. The shape formed by the string is called the *funicular curve*. Turn the image upside down to get a diagram that is similar in shape to the moment diagram.

Hang string loosely.

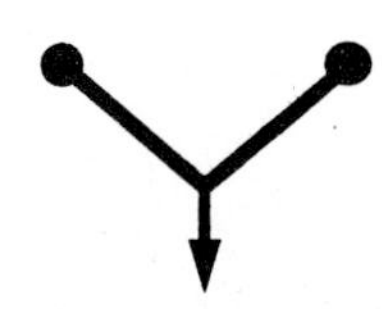

Place weight(s).

Invert the funciular curve.

To use this diagram, once you know its shape, simply calculate the moments at the transition points. These are the 3 points necessary to find the moment function. This method gives the same answer as the procedure used in Part I, but it is much quicker.

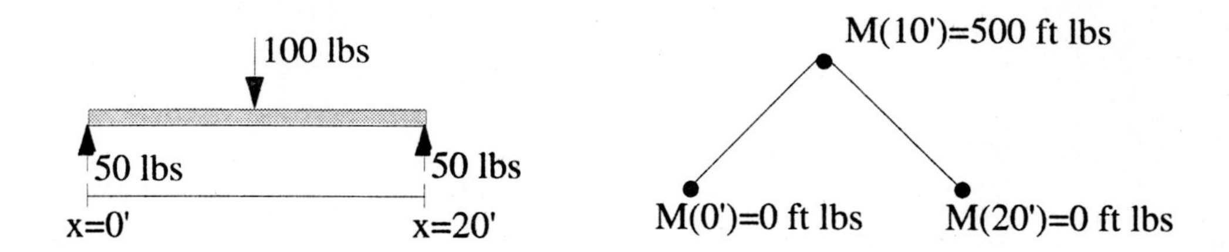

TASK 4:

Find the Moment diagram and the moment function for a 6' simply supported beam with a 180 lb load at the center.

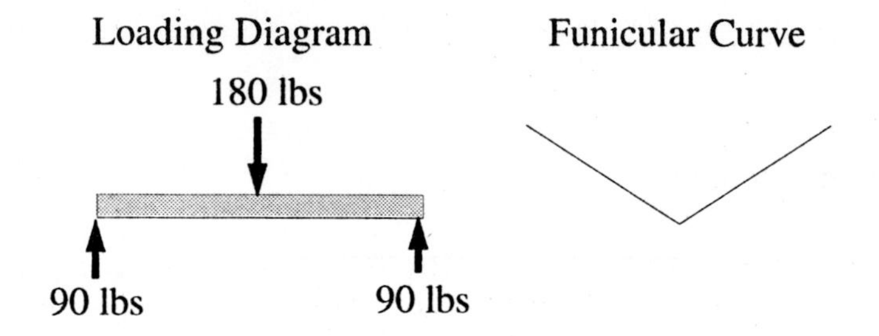

Moment calculations:

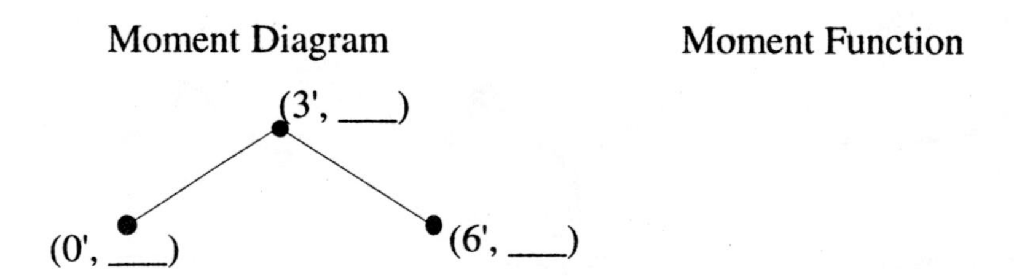

TASK 5:

Divide the 180 lb load into two equal loads and place them equally spaced along the beam. Find the moment diagram and the moment function.

Loading Diagram

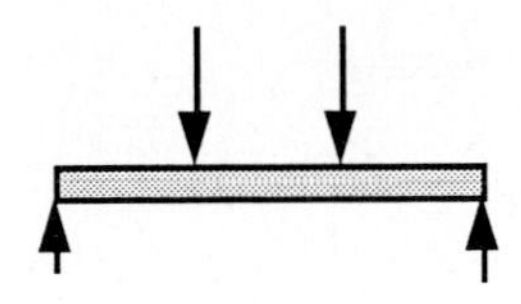

Funicular Curve

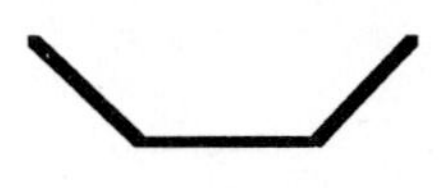

Moment Diagram

Moment Function

TASK 6:

Divide the 180 lb load into three equal loads, placed at equal intervals along the length of the beam. Draw the moment diagram and create the moment function.

Loading Diagram

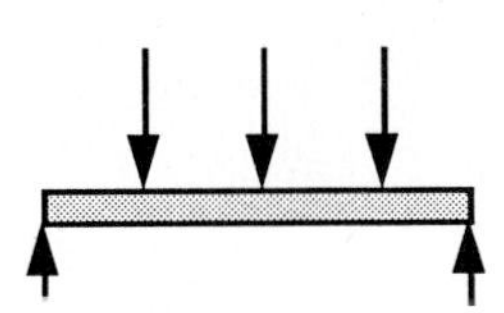

Funicular Curve

Moment Diagram

Moment Function

TASK 7:

Divide the 180 lb load equally at 1 ft intervals. What would the load be at each interval? __________

Sketch the moment diagram.

TASK 8:

What if the 180 lb force were divided into 10 equal loads placed at equal intervals along the length of the beam? Draw a rough sketch of the moment diagram.

TASK 9:

Divide the 180 lb load into 100 equal loads or 1000 equal loads equally spaced along the length of the beam. Describe what the moment diagram would look like and what the moment function would have to include.

Describe the moment diagram if the 180 lb load was uniformly distributed along the length of the beam.

The moment function is no longer piecewise-defined because it is a *continuous* curve. In order to find the equation for this curve, at least three points on the curve must be known. The endpoints of this curve are M(0')=0 and M(6')=0, but we still need a third point and have no way of getting it. An extension to this investigation develops a method for finding additional points.

Even without the moment function, the rough sketch for the moment diagram contains important information about the beam:

Where is the beam not bending at all? ______________________________

Where is the maximum bending moment? __________________________

Where is the bending moment positive? ___________________________

EXTENSION: THE MOMENT FUNCTION

Engineers, architects, mathematicians, all problem solvers often find it necessary to approach a problem from a different point of view, sometimes a more abstract point of view. In order to find the moment function for the 6 ft simply supported beam with a uniformly distributed 180 lb load we'll have to get creative.

Consider the calculated bending moments and the shear diagrams for the same loading condition. (Refer to Shear diagrams created in **Mathematics Laboratory Investigation Shear and Bending Moment I.**)

Example 1: Two 90 lb loads equally spaced along the 6 ft beam.

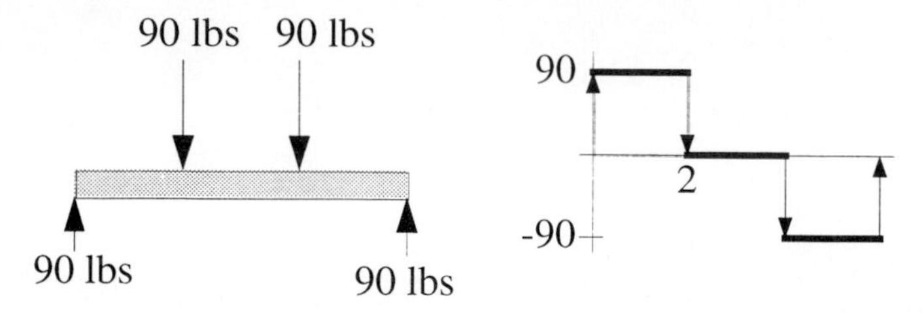

From Task 5: M(2) = (90)(2) = 180. Look at the shear diagram from a geometrical point of view. What is (90)(2)? It is (a vertical measurement)(a horizontal measurement). It's the area between the x-axis and the shear diagram between x=0 ft and x=2 ft. Will this always work? Try another x value.

From Task 5: M(4') = ____________.

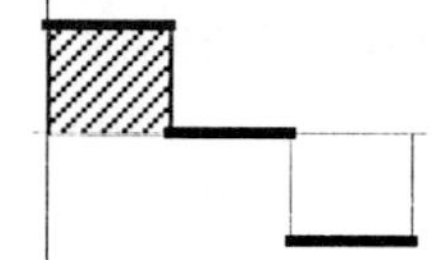

What is the area between the x-axis and the shear diagram between 0 ft and 4 ft? ____________

So far the theory works.

From Task 5: M(6') = ____________.

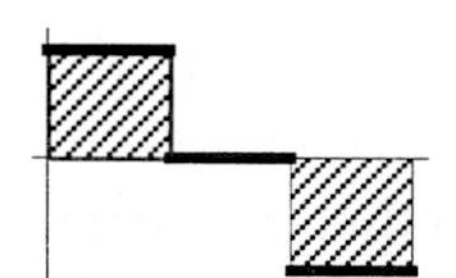

What is the area between the x-axis and the shear diagram 0 ft and 6 ft'? ____________

It seems that the theory might not hold true, however if we go back to the idea that the area of a rectangle = (vertical measurement)(horizontal measurement) we can decide that since V(4') = -90 lbs, the area between the shear diagram and the x-axis between 4 ft and 6 ft is (-90 lbs)(2 ft) = -180 ft lbs.

Using this idea, what is the area between 0 ft and 6 ft? ____________

Even though some of the area is located below the x-axis and some is located above it, the area between the x axis and the shear diagram is referred to as **the area *under* the shear diagram**.

Example 2 Three 60 lb loads equally spaced along the 6 ft beam.

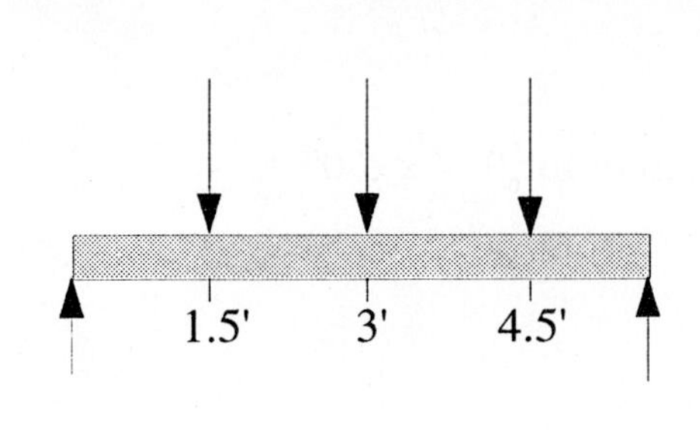

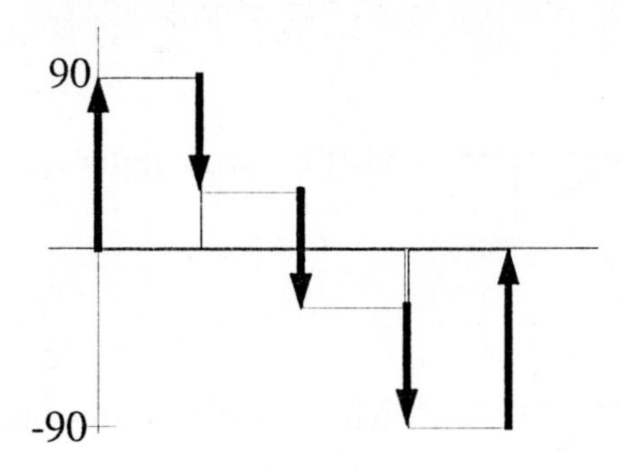

(Refer to Task 6 for the values of the bending moments.)

M(1.5') = ________
The area under the shear diagram between x=0 ft and x=1.5 ft is __________

M(3') = ________
The area under the shear diagram between x=0 ft and x=3 ft is __________

M(4.5') = ________
The area "under" the shear diagram between x=0 ft and x=4.5 ft is __________

M(6') = ________
The area "under" the shear diagram between x=0 ft and x=6 ft is __________

So it seems that the bending moment is numerically equal to the area under the shear diagram.

Now return to the case where the 180 lb force is uniformly distributed across the 6 ft beam. The shear function is V(x) = ________________

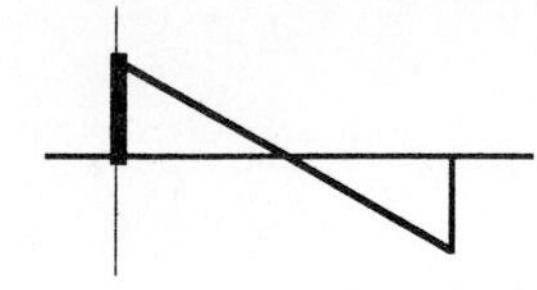

M(0') = area under the shear diagram for 0' < x < 0' = ________

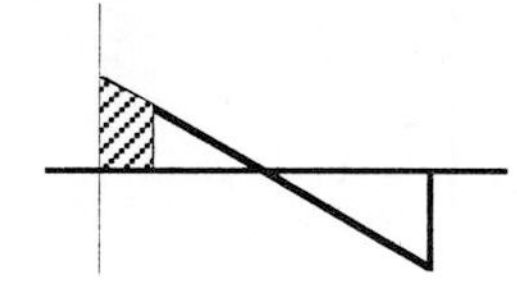

M(1') = area under the shear diagram for 0' < x < 1' = ________

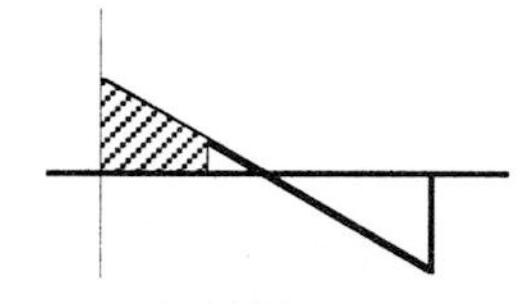

M(2') = area under the shear diagram for 0' < x < 2' = ________

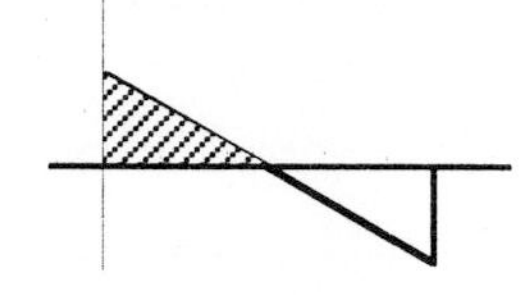

M(3') = area under the shear diagram for 0' < x < 3' = ________

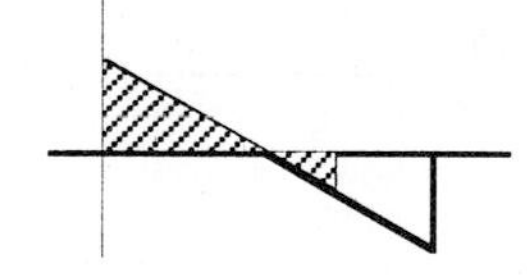

M(4') = area under the shear diagram for 0' < x < 4' = ________

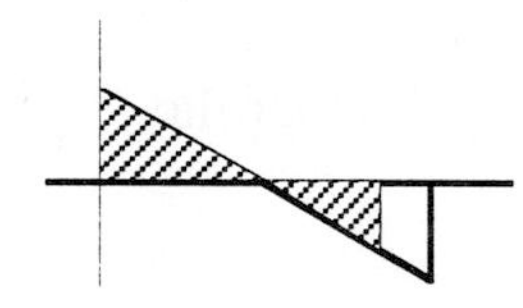

M(5') = area under the shear diagram for 0' < x < 5' = ________

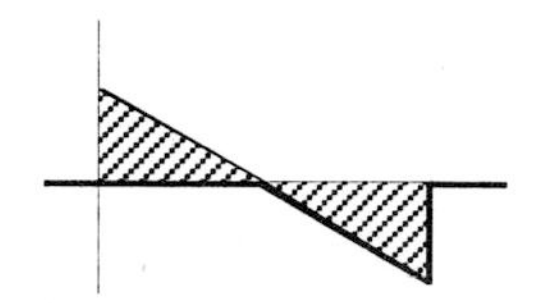

M(6') = area under the shear diagram for 0' < x < 6' = ________

Tabulate these values and use them to find the moment function for the beam.

x (in ft)	M(x) (in ft lbs)

M(x) = ________________________

Sketch a graph of the function.

With this function it is possible to find the bending moment at any point in the beam. For example, find the bending moment at 3.75 ft from the left edge. ______________

Is the bending moment positive or negative at that point? ______________

What does this mean?

Find the maximum bending moment. Where is it located?________________

How does this compare to the maximum bending moment when the 180 pound load was concentrated at the center of the beam?

MATHEMATICS LABORATORY INVESTIGATION

SOIL PRESSURE

Topic: **USE OF THE SINE FUNCTION IN FORMULAS**
Prerequisite knowledge: *Angle measure (degrees), sine function, range of sine and cosine functions*

I. INTRODUCTION

Anyone who has flown in an airplane, driven up or down a mountain road or gone diving or swimming probably has had their ears "pop" or has felt increased pressure in their ears. Coal miners or construction workers in deep mine shafts can get the bends. A container with thin or flimsy sides can carry loose dirt, but may burst if filled with water. All of these phenomena are results of horizontal pressure.

Perhaps you have tried to dig a hole only to find that at some point the deeper you dug, the more the sides of the hole collapsed. The type of soil in which you are digging plays an important role in how deep you can dig before the walls of the hole cave in. Different types of soil exert different amounts of horizontal pressure.

The effect of the horizontal pressure exerted by a particular medium, sand for example, can be demonstrated with a pile of that medium. If you pour a pile of sand onto a table and place a piece of cardboard vertically into it then gently sweep away the sand from one side of the cardboard, you will see the effect of the horizontal pressure. If you try this with different types of media, you'll see that sometimes the card will fall very quickly, while at other times the card will remain vertical even when all of the medium on one side is removed.

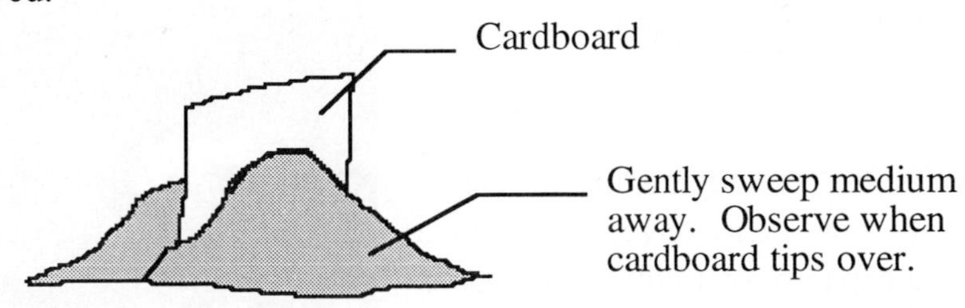
Cardboard
Gently sweep medium away. Observe when cardboard tips over.

II. THEORY

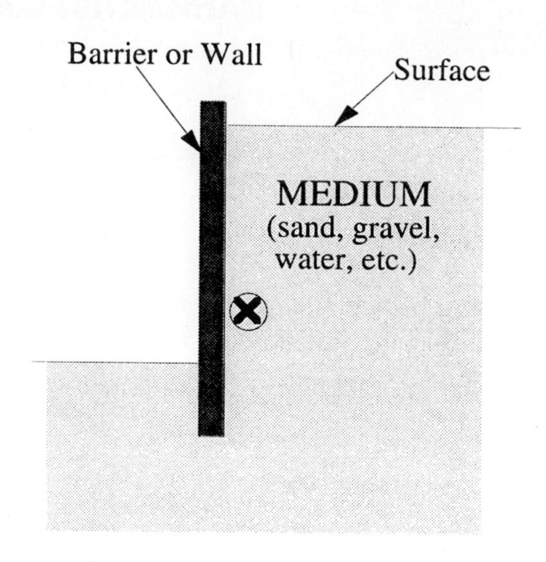

An object at the location marked with an X in the diagram would experience a *vertical pressure* exerted by the weight of the medium above it. The further down the object is, the greater the vertical pressure is. So vertical pressure is a function of both the depth and the weight of the material. You can calculate the vertical pressure P_V at specific depths below the surface by multiplying the *depth* (h, in feet) by the *unit weight of the medium* (γ, in pounds per cubic foot), so that:

$$\mathbf{P_V \text{ (at depth h)} = h\gamma}.$$

You would also feel a *horizontal pressure*. Think about a pile of sand behind a wall. If more sand is added it will slide down and press against the wall creating a component of horizontal pressure.

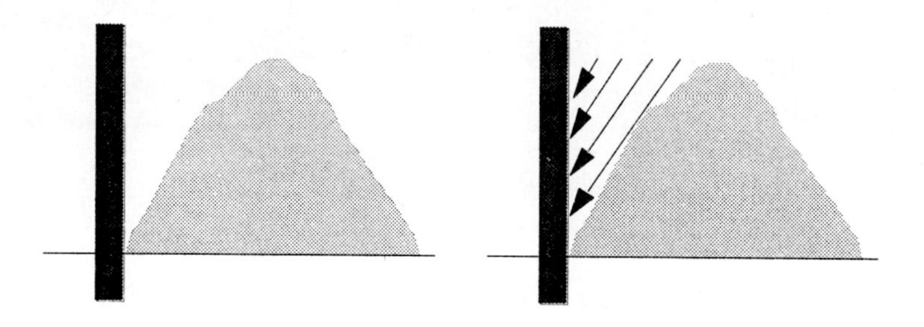

The value of horizontal pressure is a function of the depth (h, in feet), the unit weight of the medium (γ, in lbs/ft^3), and on a *coefficient of horizontal pressure* (α) which varies with the type of medium. This investigation focuses on the effect of the type of medium.

III. COEFFICIENT OF HORIZONTAL PRESSURE

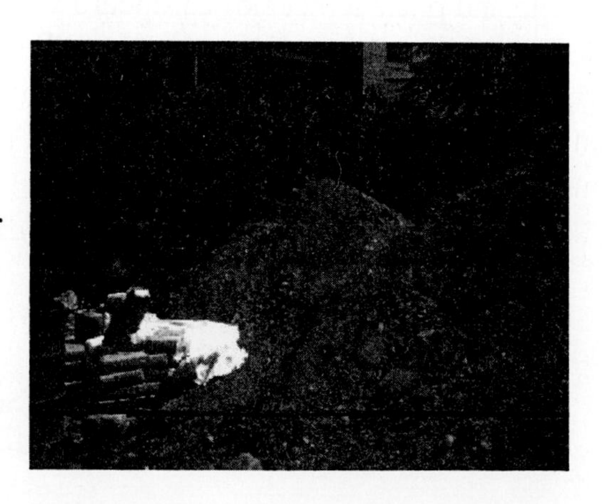

TASK 1: <u>Unit Weight (γ</u>, in pounds per cubic foot).

The value of γ for any medium is easy to find. Since it would be inconvenient to use a container with a capacity of 1 cubic foot, use a smaller container and calculate the volume of it in cubic inches. Figure out how many times you would have to fill the container to get one cubic foot of the medium.

Now fill the container with one type of medium, weigh it and calculate how much one cubic foot of that material would weigh. Record your answer in the table for results in Task 2.

Repeat the process for each of the media that you are using. Record the unit weight for each of them in the table for results.

TASK 2: Finding the *angle of repose.*

If we take a container full of a medium, turn it upside down and dump it out it forms a natural pile that looks like this:

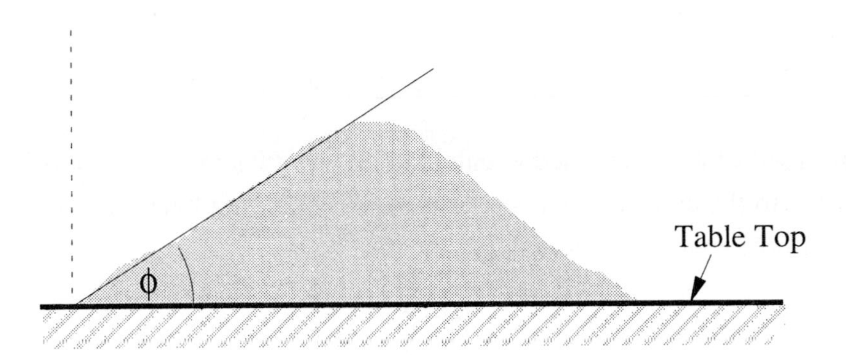

The angle that the medium makes with the table top is called the *angle of repose* or the "angle of internal friction" and is denoted by the symbol ϕ. Find the angle of repose for each of the media that you are analyzing for this investigation and record your results in the table for results.

Table for results

MEDIUM	γ (UNIT WEIGHT, lbs/ft^3) (From **Task 1**)	ϕ (ANGLE OF REPOSE, degrees) (From **Task 2**)	α (COEFFICIENT OF HORIZONTAL PRESSURE) (From **Task 3**)
Sand			
Gravel			

What is the range of possible values of ϕ for <u>any</u> medium? Why?

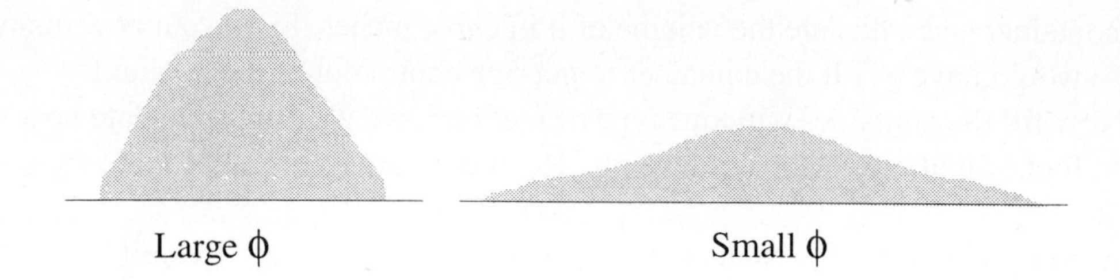

If the value of ϕ is high, would you expect more or less horizontal pressure than if the value of ϕ is low? Why?

TASK 3: Coefficient of Horizontal Pressure (active case)

The angle of repose is used to calculate the *coefficient of horizontal pressure*, denoted by α. In the case where a barrier is actively holding back a medium, the coefficient of horizontal pressure is

$$\alpha = \frac{1 - \sin \phi}{1 + \sin \phi}$$

Calculate the coefficient of horizontal pressure for each of the media and record your results in the table on the previous page.

What is the range of possible values of α for <u>any</u> medium? Why?

Experiment with different possible values of ϕ and discuss the relationship between the angle of repose ϕ for a medium and the coefficient of horizontal pressure α for the same medium. For example, if ϕ is large, will α also be large? Write your conclusions here. How do the characteristics of the sine function explain how α and ϕ are related?

IV. TWO MEDIA OF SPECIAL INTEREST

What if our medium were water? What would the values of ϕ and α be?

Why?____

____

____

____

What if our medium was very stiff clay? What would the values of ϕ and α be?

Why?____

____

____

____

V. CALCULATING AND PLOTTING THE HORIZONTAL PRESSURE

The actual horizontal pressure at any depth can be calculated using $P_H = \gamma h \alpha$. (Notice that $P_H = \alpha$ times P_V or αP_V. That is why α is called the coefficient of horizontal pressure.)

Consider the horizontal pressure behind a retaining wall 23 ft in height that is being designed to support a 20 ft deep cut with the base of the wall located 4 ft below grade. The soil behind the wall has a unit weight of 125 lbs per cubic foot and an angle of repose of 35°.

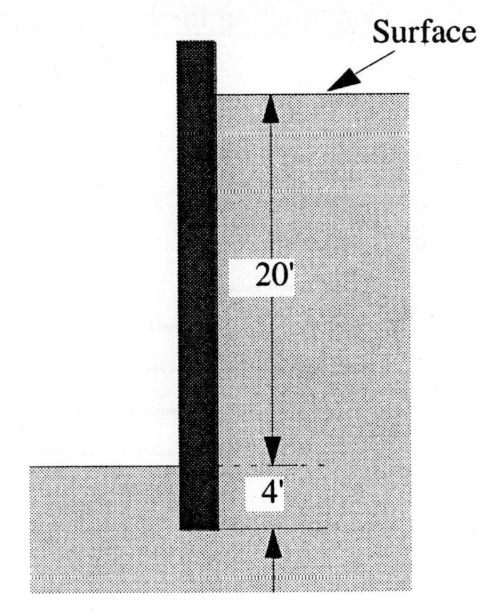

TASK 4: Calculating values.

Fill in the table with the values of P_H at the depths indicated. Using a spreadsheet or list feature on a calculator will make this task less tedious.)

DEPTH FROM SURFACE (in ft)	HORIZONTAL PRESSURE (in lbs/ft^3)
0 ft	
3 ft	
6 ft	
9 ft	
12 ft	
15 ft	
18 ft	
21 ft	
24 ft	

TASK 5: Constructing the horizontal pressure diagram.

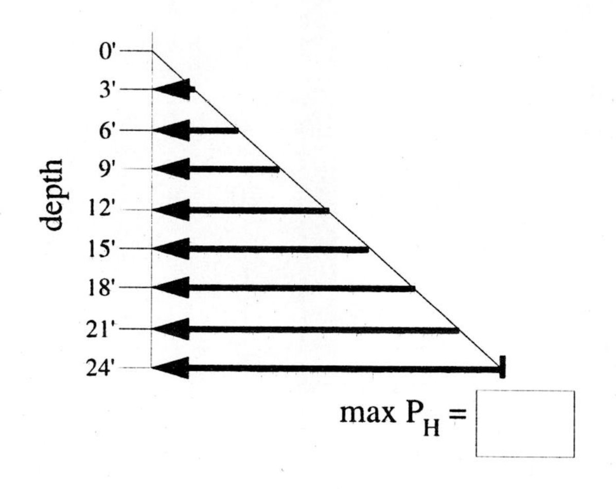

It is often important or helpful to have several methods of organizing numerical information. A *horizontal pressure diagram* shows the pressure at various depths as horizontal vectors.

This diagram illustrates that the greatest horizontal pressure occurs at the base of the wall and decreases to zero at the surface. Fill in the maximum horizontal pressure in the space provided in the diagram.

TASK 6: Plotting values.

Another method of organizing numerical information is to create a graph by plotting points on a rectangular coordinate system and finding a curve of best fit. Plot the information from Task 4. (Use your calculator or spreadsheet to do this if possible.)

Notice that these points all line up along a straight line. If you find the equation for that line you can use it to find the horizontal pressure at any depth. For example, what is the horizontal pressure at a depth of 16.5 ft below the surface?

Draw a sketch of the graph here. Be sure to label the axes.

TASK 7: Another example.

Choose one of the media measured in part II and create a table, a horizontal pressure diagram and a graph with the line of best fit.

Table:

DEPTH FROM SURFACE (in ft)	HORIZONTAL PRESSURE (in lbs/ft^3)
0 ft	
3 ft	
6 ft	
9 ft	
12 ft	
15 ft	
18 ft	
21 ft	
24 ft	

Horizontal Pressure Diagram:

Graph:

TASK 8: Comparing vertical pressure and horizontal pressure.

Add a column with vertical pressure (see p. 2) to the table you created in Task 7. Which pressure is greater, horizontal or vertical? Why do you think this is true? Does it seem realistic?

TASK 9: An interesting case.

What if the wall is a dam and is holding back water? Fill in the table below and plot the horizontal pressure diagram and the line graph. The unit weight of water is 62.4 lbs/ft^3.

DEPTH FROM SURFACE	VERTICAL PRESSURE (see p. 2)	HORIZONTAL PRESSURE

For water the horizontal pressure is equal to the vertical pressure! This fact is important to scuba divers as well as to designers of submarines, oil platforms, etc.

Explain why this is true using the fact that $\alpha = \dfrac{1 - \sin\phi}{1 + \sin\phi}$

VI. GENERALIZE

Many formulas contain trigonometric functions. The following problems are examples from several different fields.

If we consider an angle θ that is between 0° and 90° inclusive, what is the possible range of values of the $\sin\theta$?

What is the possible range of values of the $\cos\theta$?

A. The maximum height a projectile fired at an inclination of θ to the horizontal with an initial speed of v_0 is given by

$$H = \frac{v_0^2 \sin\theta}{2g}$$

where g is the acceleration due to gravity.

If $v_0 = 120$ ft/sec and $g = 32$ ft/sec^2

(1) What is the maximum height a projectile reaches if it is fired at an angle of 30° from the horizontal?

(2) What is the maximum height if the angle $\theta = 90^\circ$?

(3) What is the maximum height if the angle $\theta = 0^\circ$?

B. The length of time t that a projectile is in flight is given by $t = \dfrac{v_0 \sin\theta}{g}$.

If $v_0 = 120$ ft/sec and $g = 32$ ft/sec^2, what angle of inclination θ will maximize the time of flight?

C. The tension T at any point in a cable supporting a distributed load is given by

$$T = \frac{T_0}{\cos \theta}$$

where T_0 is the tension where the cable is horizontal and θ is the angle between the cable and horizontal at any point.

For what angle θ will T be minimized?

D. The power P of an electric circuit is given by $P = P_a \cos \theta$ where P_a is the apparent power and θ is the impedance phase angle.

(1) If $P_a = 12$ watts, what is the <u>maximum</u> value of P? At what value of θ does this occur?

(2) If $P_a = 12$ watts, what is the <u>minimum</u> value of P? At what value of θ does this occur?

There are many formulas involving trigonometric functions that are not limited to angles between 0° and 90° or that use radian measure. The important thing to remember is that the sine and cosine functions have a specific range no matter what the angle or argument of the function is.

The graphs of the sine and cosine functions are:

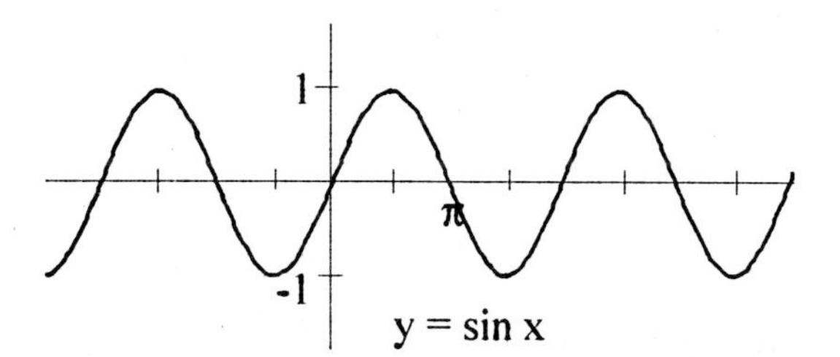

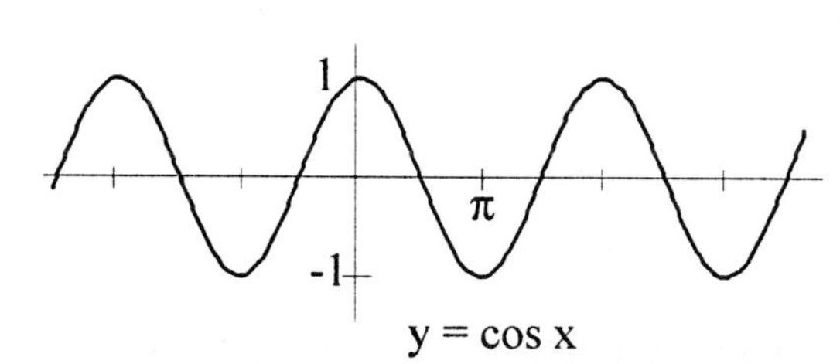

What is the range of possible values for the sine function?

What is the range of possible values for the cosine function?

E. The average temperature in Fairbanks, Alaska is given by

$$T = 37 \sin\left(\frac{2\pi}{365}(d - 101)\right) + 25$$

where T is the temperature in °F on day d of the year.

1) What is the range of possible values of $\sin\left(\frac{2\pi}{365}(d - 101)\right)$?

2) What is the maximum average temperature possible using the formula?

3) What is the minimum?

F. The formula that describes the motion of a spring that has been stretched 20 cm and released (ignoring friction) is $d = 20 \cos(\omega t)$ where d is the displacement from the position of the unstretched spring after t seconds.

(1) What is the range of possible values of $\cos(\omega t)$?

(2) What is the maximum displacement possible using this formula?

(3) What is the minimum?

G. The voltage V at time t in a particular electric circuit is given by $V = 25 \sin 2t$.

(1) What is the range of possible values of $\sin 2t$?

(2) What is the maximum voltage possible using this formula?

(3) What is the minimum?

V. EXTENSION 1: SURCHARGE LOADS

If a parking lot were built on the surface next to the wall we were considering in Part V, would the parked cars have an impact on the horizontal pressure on the wall?

TASK 9:

Consider the effect of a vehicle that is parked next to the wall. The Jeep Cherokee shown is approximately 5 ft high, 14 ft long, 6 ft wide and weighs approximately 3000 lbs.

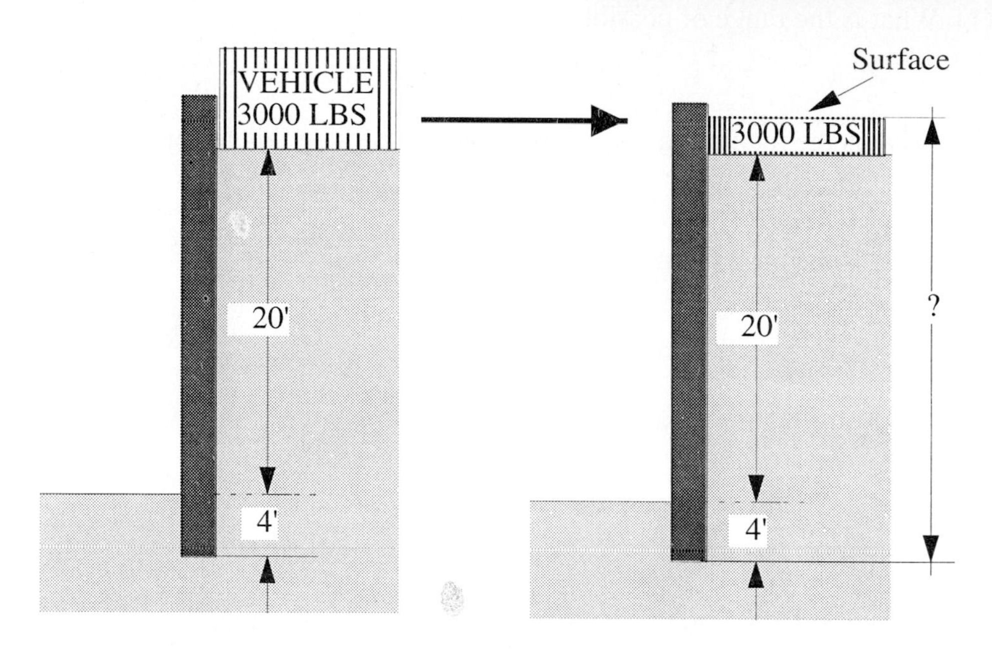

To find the effect of the vehicle on the soil pressure, convert the Jeep into an equivalent amount of "imaginary soil": The soil under consideration has a unit weight of 125 lbs per cubic foot. How deep would a rectangular solid of soil be if it weighed 3000 lbs and had a 14 ft by 6 ft base?

With the Jeep converted to the equivalent amount of soil, how far above the base of the wall will the new surface be? ______________________________

Fill in the table below paying careful attention to the depth.

Depth from Surface	Horizontal Pressure Due To Soil
0 ft	
3 ft	
6 ft	
9 ft	
12 ft	
15 ft	
18 ft	
21 ft	
24 ft	

Has the horizontal pressure at the base of the wall changed because of the parked vehicle? Why or why not?

This type of load is called a *surcharge load.* Create the horizontal pressure diagram for the wall. How does it differ from the one in Task 5?

Create a line graph for these values. How does it differ from the one created in Task 6?

VIII. EXTENSION 2: COEFFICIENT OF HORIZONTAL PRESSURE (PASSIVE CASE)

This investigation has been concerned with the case of a medium actively applying pressure to a barrier. The situation is different if the barrier is applying pressure to the medium. As an example of this situation think about opening your outside door after a snowstorm. As you push the door, it is the "barrier" and it is applying pressure to the snow. In this case the medium is considered passive.

The coefficient of horizontal pressure for the passive case also depends on the angle of repose. It can be calculated with

$$\alpha = \frac{1 + \sin \phi}{1 - \sin \phi}$$

Calculate the coefficient of passive horizontal pressure for each of the media used in this investigation. Compare them to the coefficient of active horizontal pressure. How are they related?

There are two cases of special interest that are worth time and discussion:

1) Consider the passive horizontal pressure for water. ($\theta = 0°$)

2) Consider the passive horizontal pressure for a very stiff clay. ($\theta = 90°$)

MATHEMATICS LABORATORY INVESTIGATION

STRENGTH OF MATERIALS

Topic: **INTERPRETATION of SLOPE**
Prerequisite knowledge: *Graphing (scatterplots)*
Equipment needed: Rubber band, stand or hook, hanger and weights, millimeter scale, calipers, a piece of extruded polyethylene (~1/8" diameter)

I. INTRODUCTION

In some engineering applications, the importance of material strength is obvious, as when wood is used to frame a house or a high-rise building is constructed of steel-reinforced concrete. Even the foam plastic used in automobile bumper cores clearly has strength requirements. Less obvious is the need for a knowledge of the strength of thin plastic wrap used to package the chicken you might buy at your supermarket; but the manufacturer of that plastic wrap must take the strength of the plastic into account in order to produce a package that will wrap snugly without tearing.

More precisely, the *tensile strength* of a substance measures its ability to withstand tension ("pulling") forces, such as the force acting on a bridge's supporting cables. (The opposite of this would be *compressive* strength.) A stronger material will respond to a given "stretch" with a greater force than a weaker one.

In order to allow for direct strength comparisons between different materials, the force necessary to stretch a unit cross-sectional area of a material specimen by a certain fractional amount of its length is measured, and is called the *modulus of elasticity*. (It is also sometimes called *Young's modulus,* after the person who first defined it.) The fractional stretch of the material is referred to as *strain*, and the response force per unit area is called *stress*. Hence the modulus of elasticity **E** is the ratio of stress σ to strain ε, or $E = \sigma/\varepsilon$. The value of E is then a direct measure of the material's strength, and allows direct comparison of the ability of different materials to resist stretching.

II. THE BEHAVIOR OF A MATERIAL UNDER TENSION

TASK 1: If a length of extruded polyethylene is available, you can get some sense of how an elastic material behaves when subject to a tensile (stretching) force. Get a good grip on both ends of the plastic (you may have to protect your hands), and pull on it with a gradually increasing force. Continue increasing the force of your pull until it stretches noticeably.

How did the appearance of the plastic change as it stretched?

Describe any changes you noticed in the plastic's resistance to stretching as you continued to pull on it.

III. MEASURING MODULUS OF ELASTICITY

Although it is not practical to make strength measurements on steel and most other structural materials without sophisticated equipment, you can determine the modulus of elasticity for a rubber band.

TASK 2: Suspend a thick rubber band from the spring scale as shown in the figure at the right. At the bottom of the rubber band, attach a hanger to which you can add increasing amounts of weight so that you can measure the stretch of the rubber band. Mark two horizontal lines near opposite ends of the rubber band in order to make it easier to measure precise changes in length.

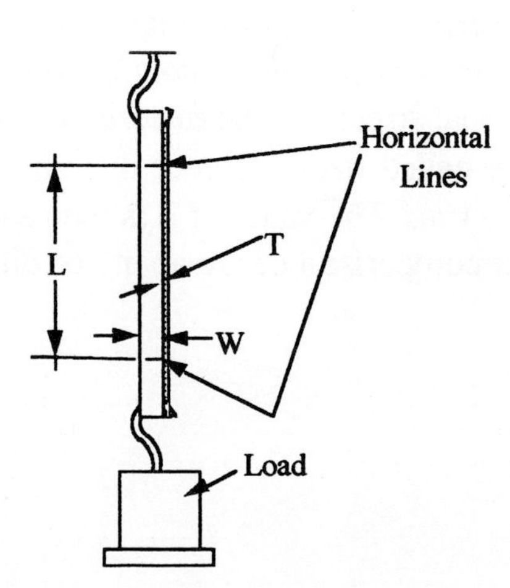

Add weights to the hanger and record data that will allow you to determine the longitudinal strain ε (the change in the length between marks on the rubber band divided by the original length) as well as the stress (force divided by the area of a cross-section of the rubber band perpendicular to the direction of stretch) exerted by the rubber band for each weight total.

Explain below what measurements you think are necessary in order to obtain the required data, and show a complete table of all data including the calculated strain and stress values for each total weight; the LIST statistics on a TI calculator may be helpful here. (Include appropriate and consistent units for all quantities.) You may have to try this more than once before realizing what data must be measured and recorded.

TASK 3: Now construct a well-labeled scatterplot of stress vs. strain for the rubber band, including units where appropriate. (Remember that stress measures force per unit area, and strain measures a fractional change in length.)

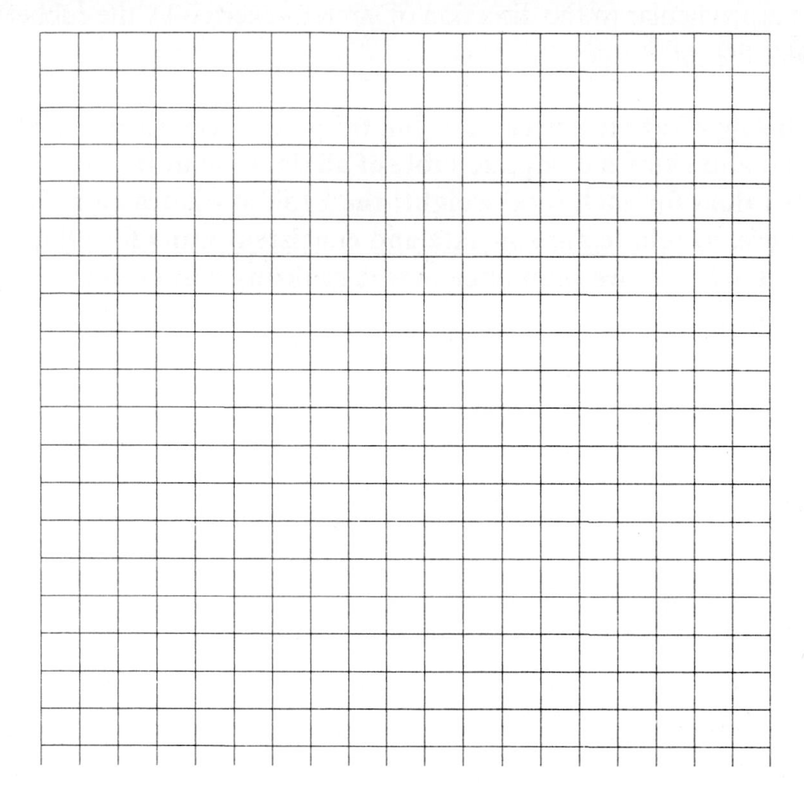

Until the weight gets too heavy, the points on your graph should appear to fall approximately in a straight line. (After that the graph may change to a line with a different slope, or even a curve.) Draw in your best "eyeball" estimate of a line that fits the initial linear part of the data. **You should draw this line through the origin; explain why this should be so.**

TASK 4: The modulus of elasticity of the rubber band is equal to the slope of the stress-strain line on your graph. **Pick any two points on your line (not data points) and use them to calculate the slope, which should equal the rubber band's modulus of elasticity. (Again, think carefully about the appropriate units for your result.)**

TASK 5: As an alternative to calculating a single slope value, **calculate the separate values for slope (and thus modulus of elasticity) from the origin to <u>each</u> of the points on the initial straight-line portion of the graph, and determine their average.** (Note: Linear regression methods are not suggested here, as they may not produce a line that passes through the origin.)

TASK 6: If you took into account the reduction in cross-section of the rubber band as it stretches, then the stress values you calculated up to this point are referred to as *true stress*. However, materials engineers in some cases ignore this reduction and report stress without respect to change in cross-section; in this case, the stress values are referred to as *engineering stress*. **Which type of stress did you use in your analysis?**

How do you think your graph and your value for modulus of elasticity would have been different if you had used the other type of stress for your calculations?

IV. MODULUS OF ELASTICITY FOR OTHER MATERIALS

Many materials have the ability to either stretch and return to their original shape, or stretch and remain deformed after the load is released. Rubber is said to be *elastic*: remove the load and it snaps back. On the other hand, clay is very *malleable*: it deforms easily in compression without breaking apart. Some materials, such as glass, ceramics, and concrete, don't do either very well. Materials that do both--many plastics and metals, for example--are desirable for product design applications. Butter dish tops snap into place and deform to provide airtight containers; an airplane wing is formed into a foil shape for lift, and then bounces up and down slightly during flight without snapping off (a very fortunate property!). An airplane once hit the Empire State Building near the top, causing it to sway more than 10 feet. Imagine the consequences if the material used for the building were not able to deform elastically.

The elastic characteristics of many materials are linear; i.e., material stress is a linear function of deformation (strain), at least at relatively low strain levels. The graphs below show stress-strain plots for two types of plastics, based on data from a Tinius-Olsen universal testing machine, a drawing of which is shown at the right. Notice that the strain values are given units of inches of deformation divided by original length (in inches), which actually results in a unitless ratio: strain measures fractional deformation. Here, the height of the horizontal part of the graph gives a relative measure of the amount of force required to permanently deform the piece compared to other materials.

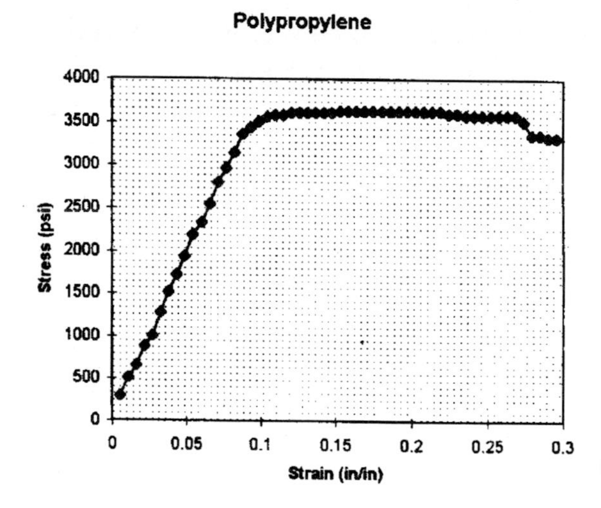

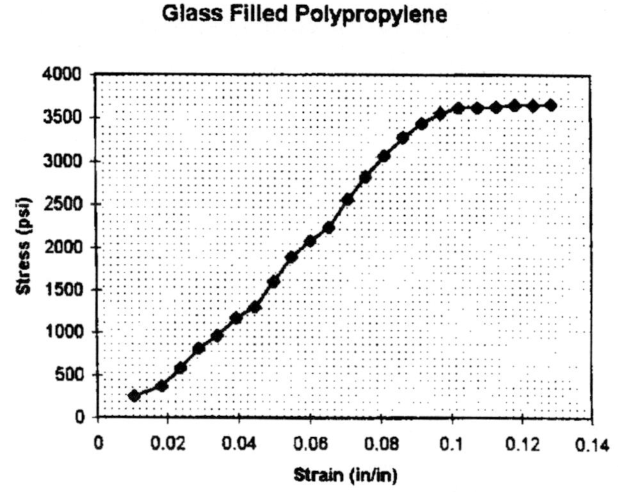

TASK 7: **Calculate the modulus of elasticity for each plastic, again using only the initial linear part of the graph (taking into account the differing scales of the two graphs). Which is stronger in this initial linear, or *elastic*, range?**

Based on your observations of what happened as you stretched the polyethylene, and assuming that a similar process happened with the plastic specimens that produced the graphs on the previous page, **see if you can explain the shapes of the graphs by relating them to the behavior of the polyethylene.**

V. SLOPE AS A RATE OF CHANGE

Since slope is calculated from the ratio of two quantities, it can also be interpreted as a measure of the way in which *changes* in those quantities are related.

TASK 8: The graph at the right shows the distance traveled by a car on a highway (measured in miles) as a function of time (measured in hours). **The slope of the graph indicates that, in most states, this car would be breaking the law. Explain why this is true by calculating the slope, along with its units, and interpreting its meaning.**

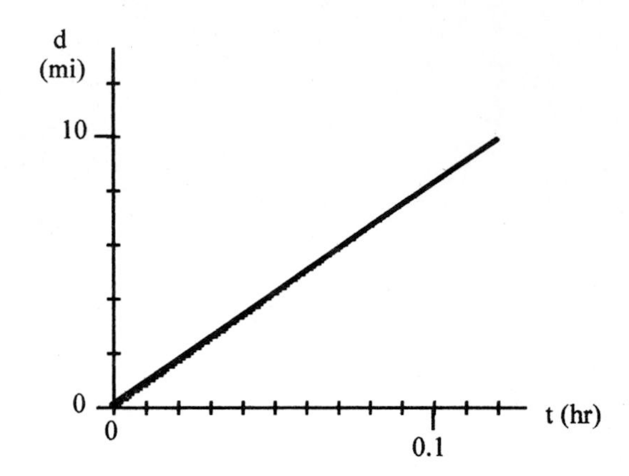

This is an example of the fact that the slope of a graph measures the *rate of change* of the dependent variable with respect to the independent variable. If the graph is a straight line, then the slope is constant, and it follows that the rate is also constant. **In the distance-time graph above, what is a common name for the constant rate of change that is measured by the slope of the graph?**

The symbol delta Δ (capital D in the Greek alphabet) is often used to indicate a *change*, or *increment*, or *interval* between values in a graph or table. Hence **Δd** refers to a distance interval, and **Δt** to a time interval. Then the constant slope of the distance-time graph above can be symbolized **Δd/Δt**. This fractional "delta" notation is a common way of symbolizing a rate of change.

TASK 9: Each of the graphs on this page shows a linear relationship between two quantities. In each case, the slope of the graph represents a third quantity that is the rate of change of the first two, and the name of the slope quantity is given. **For each case, write the "delta" ratio that is equal to the named slope quantity, and calculate its value (including units). In some cases, there is a special name for the unit of the slope quantity--if you know it, include it as well.**

A. Charge q on a capacitor in an electrical circuit as a function of time t:

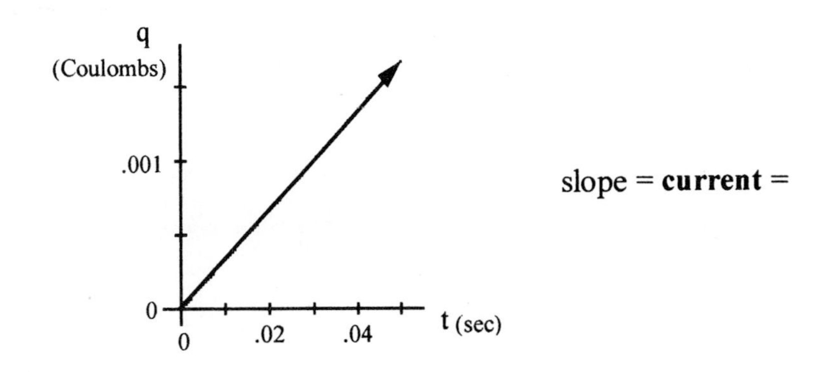

slope = **current** =

B. Temperature T in a wall as a function of distance x from one surface:

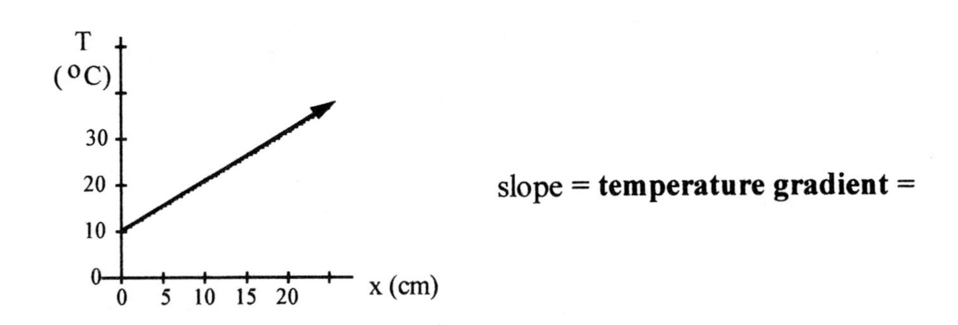

slope = **temperature gradient** =

C. Bending moment M of a beam as a function of distance x from an end:

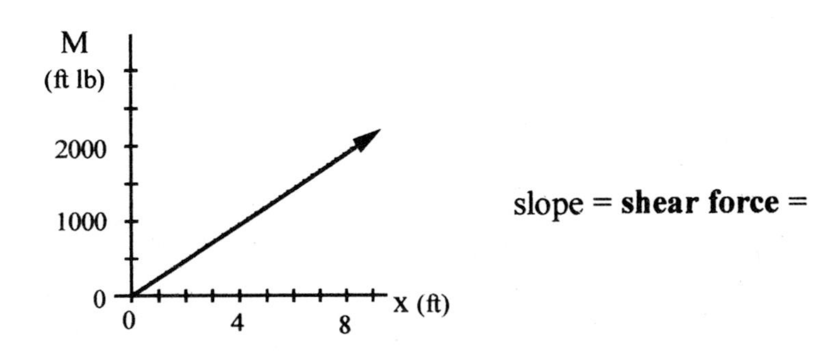

slope = **shear force** =

EXTENSIONS

1. You may have noticed that the modulus of elasticity for the plastics discussed in part III was calculated in pounds per square inch (psi), but that your result for the rubber band in part II most likely had SI (metric) units. By converting units from one system to the other, determine how many times stronger the plastics are in comparison to the rubber band.

2. Describe some other situations in which the slope of the graph of two real quantities represents a third measurable quantity.

MATHEMATICS LABORATORY INVESTIGATION

TEMPERATURE SENSING DIODE

Topic: **LINEAR MODELING,** with **GRAPHING ACCURACY, PREDICTION**
Prerequisite knowledge: *Plotting Points, Equations and Graphs of Straight Lines*

Equipment: Celsius scale thermometer, diode, 10K Ohm resistor, 9V battery, voltmeter, connecting wires, diode sheath or plastic straw, large Styrofoam cup, water, ice.

I. INTRODUCTION

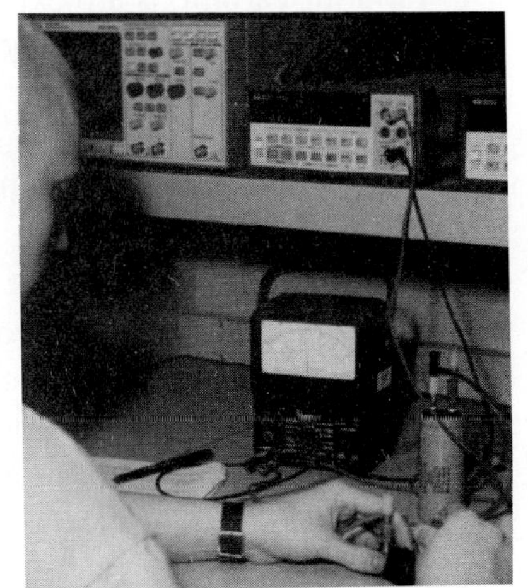

A silicon diode responds to changes in temperature. The voltage across the diode changes with the heat level the diode detects; the higher the temperature, T, the lower the voltage, V. The voltage ***varies linearly*** with respect to heat. That is, a specific change in temperature, ΔT, always produces the same amount of change in voltage, ΔV, no matter what the initial temperature was.

In this investigation you will collect experimental data and represent it in a linear model. Every physical experiment involves errors. You will learn how to reduce errors and see how they influence the linear mathematical relationship of the situation.

Linear relationships are very common in engineering applications. Even when relationships among quantities are not linear, the engineer will often make simplifying assumptions to a problem so that a linear relationship is applicable.

II. DATA COLLECTION

TASK 1:

The purpose of this task is to collect temperature and voltage data with reduced measurement errors.

Using an ordinary Celsius scale thermometer, measure the air temperature in your work area and record the result as T_1 in the table below. Be sure nothing is touching the bulb end of the thermometer and that you hold it still until the temperature reading stabilizes.

Connect the diode, resistor, battery, and voltmeter as shown in the figure above and demonstrated by your instructor. There is no danger of electrical shock.

To measure the air temperature using the diode, suspend the diode in air. Again, there is no danger but to get an accurate reading be sure not to touch both connecting wires. Record the voltmeter readout in the table as V_1.

Temperature °C	Voltage mV (millivolts)
T_1=	V_1=
T_2=	V_2=

Air temperature is stable but skin temperature measurements will change according to how you hold the diode or thermometer and how much pressure you apply. When such random variations are expected, a standard experimental method is to take several measurements and average them. This is one error reduction technique.

Follow these instructions carefully to take your skin surface temperature using the diode. One person should take all skin temperature readings with both the thermometer and the diode.

Hold the diode between your thumb and forefinger. Again, to get an accurate reading, be sure your fingers do not touch uninsulated portions of <u>both</u> connecting wires at the same time. It will be difficult not to touch one wire; that is OK. Wait until the voltage reading stabilizes then record the voltage as reading 1.

Take and record three more voltage readings. Be sure you let the voltmeter stabilize for each reading.

Calculate the average voltage reading.

VOLTAGE

reading 1	reading 2	reading 3	reading 4	AVERAGE
				$V_2 =$

Record the average voltmeter value as V_2 in the first table. Record the same decimal place accuracy as the stable digits of the voltmeter.

Hold the thermometer bulb between your thumb and forefinger and take four readings being sure the thermometer stabilizes on each reading. You should hold the thermometer bulb for two to three minutes for each reading. Calculate the average temperature.

TEMPERATURE

reading 1	reading 2	reading 3	reading 4	AVERAGE
				$T_2 =$

Record the average temperature reading as T_2 in the first table. What decimal place accuracy is appropriate to record? You probably can't read the thermometer to any more than one decimal place accuracy.

The first table now has two data points:

$$(T_1, V_1) \qquad \text{and} \qquad (T_2, V_2)$$

which correspond to room temperature and skin temperature measurements, respectively.

III. PREPARATION

Later, you will be asked to measure the temperature of ice water. It may take up to 15 minutes for the thermometer reading to stabilize. You can save time by preparing for that experiment now.

Stir the water and ice.

Place the thermometer in the ice water mixture.

Carefully place the diode in its protective sheath.

Place the protected diode in the ice water being careful to keep water from going over the top of the protective covering.

After 10 minutes check the thermometer and voltmeter readings. After 5 more minutes check them again. If they have not changed, record the results as reading 1 in the TEMPERATURE and VOLTAGE tables under TASK 4.

Repeat this procedure until you have 4 readings for temperature and voltage. You can collect this data while working on TASKS 2 and 3 below.

IV. DATA REPRESENTATION

TASK 2:

The purpose of this task is to represent the data from TASK 1 with a mathematical model.

Carefully draw and label a horizontal T axis on the grid below. Scale the T axis from 0 to 100 degrees Celsius.

Carefully draw and label a vertical V axis and scale it from 0 to 1000 mV (millivolts.)

With equal care, plot the two data points on the grid.

Physically, the relationship between heat and voltage is linear; mathematically, two points determine a straight line. Using a pencil and straight edge or ruler lightly draw a straight line through these two points. Draw the line so that it crosses the V axis. It is important to be as precise as you can.

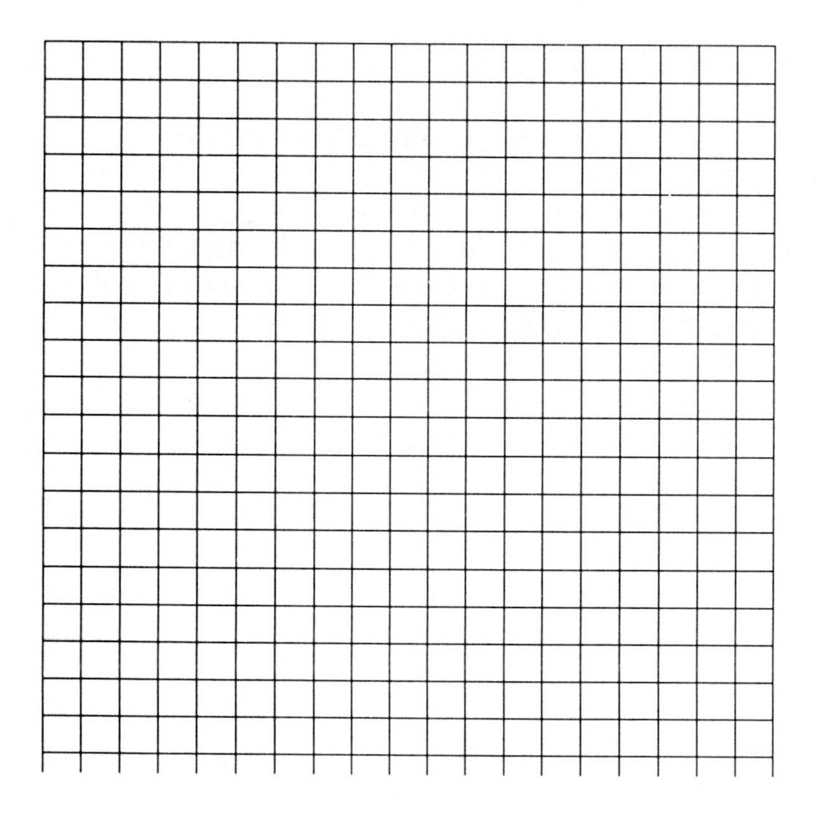

TASK 3:

The purpose of this task is to refine the data representation you just constructed in TASK 2. This will reduce errors in how you represented the data.

Use the two data points to algebraically determine the slope of the line you just sketched in TASK 2.

m = ______

Now use one of the data points and the calculated slope, m, to determine algebraically the V-intercept of the line.

$\boldsymbol{b}$ = ______

The calculated value of $\boldsymbol{b}$ provides a check on how accurately you plotted the data points and how accurately you drew the line. Describe any difference between the calculated V-intercept, $\boldsymbol{b}$, and the position your line in TASK 2 crosses the V-axis.

If the V-intercept as drawn does not look accurate compared to the calculated value of $\boldsymbol{b}$, try adjusting your straight edge to better align it with the plotted data points and the calculated V-intercept, $\boldsymbol{b}$. If your straight edge still doesn't look like it hits the data points and the calculated V-intercept, examine how accurately you plotted the data points and the intercept. You may have to refine those plotted points so they still accurately represent the data and the calculated V-intercept. **Do not** fake the data or the plotting. You are using the mathematics of slope and intercept to help reduce plotting and drawing errors.

Draw in a refined straight line on the grid in TASK 2 that looks like a good fit to the data and the calculated V-intercept. What you have accomplished is a good representation of the actual data.

V. MEASUREMENT ERRORS VS. PLOTTING ERRORS

Data points were averaged to reduce errors in taking measurements. The algebraically accurate V-intercept ***b*** was used to reduce plotting errors. Another source of error is in the measurement devices. How accurate are the diode and thermometer? Water and ice in an equilibrium solution is, by definition, at 0° Celsius. This will be used to check accuracy of the thermometer and the diode.

TASK 4:

The purpose of this task is to get some indication of experimental errors due to inaccuracy in the measurement devices. Some error may also be due to uncontrolled conditions and insufficient care in conducting the experiment.

Continue recording temperature and voltage readings for ice water as described in section III.

When you have collected 4 readings for temperature and voltage, calculate their averages. If you have time while waiting for the temperature readings to stabilize, review your previous work for possible sources of error and then compare your data and graphs with those of other teams. In real life situations you will work in teams, review your own results and consult with others who have done similar work before presenting and defending your results to your supervisor. If your team results differ from others in ways you cannot explain, consult with your instructor/supervisor.

TEMPERATURE

reading 1	reading 2	reading 3	reading 4	AVERAGE
				T_3=

VOLTAGE

reading 1	reading 2	reading 3	reading 4	AVERAGE
				V_3=

The data pair (T_3, V_3) should be the same as the pair $(0, \boldsymbol{b})$. Under ideal conditions any discrepancy is due to measurement error. Plot the point (T_3, V_3) on the previous grid.

Explain the significance of any difference between the point (T_3, V_3) and the refined V-intercept.

TASK 5:

The purpose of this task is to create a mathematical model based on the best available information. You have taken steps to assure the data collection process was accurate. You have refined your representation of the data. You have some sense of how accurate the measurement devices are. Your model should be as good as possible.

Remember, two points determine a line. With three points, there may be no line that fits them exactly.

On the grid below label the Temperature and Voltage axes as before. Plot the three data points $(T_1, V_1), (T_2, V_2) and (T_3, V_3)$ again. Remember from your work above to be VERY careful how you plot the points.

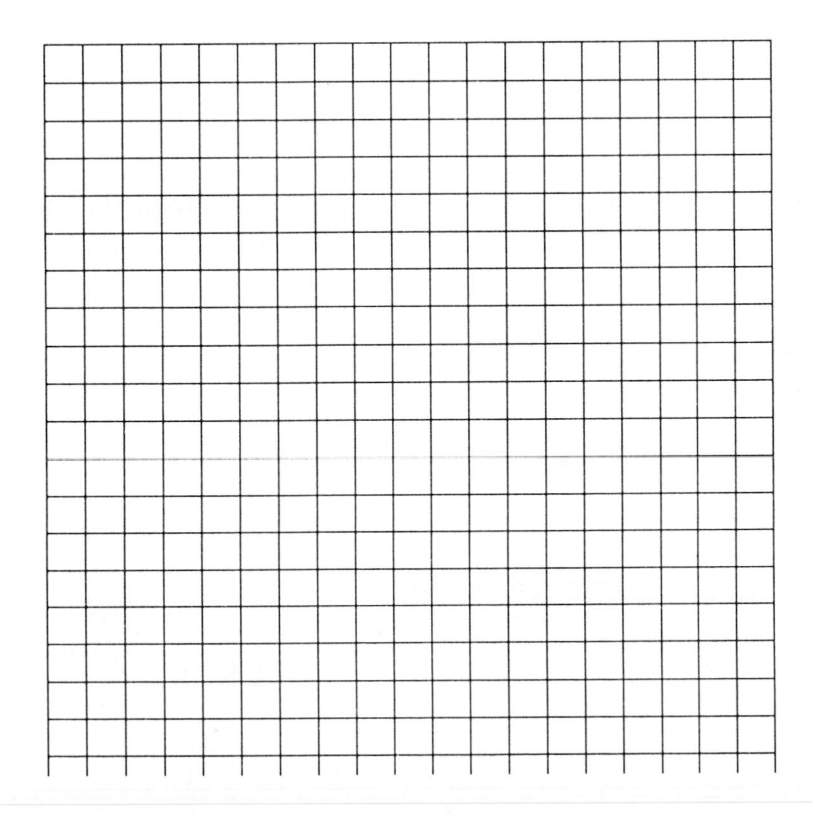

Draw the straight line that appears to best fit the three data points. The line may not "hit" any of the points but it should come as close as possible to all of them.

Determine the equation of the line from the graph.

Equation:______________________________

If you have access to a sophisticated calculator, a spread sheet or mathematical software, it probably has the capability of determining the equation of the line of best fit. Use it to find the actual equation of best fit.

Equation:______________________________

Graph this equation on the grid above.

Compare the slope and V-intercept of this equation and its graph to the one you just determined.

This will give you some indication of how the slope and V-intercept affect the appearance of the lines. It will also give you some indication of how good your judgments were in estimating the line of best fit.

VI. DATA PREDICTION AND CHECKING

TASK 6:

The purpose of this task is to use the mathematical model you created to predict the voltage outcome of a temperature measurement. It serves as a check on the accuracy of the best fit line.

Use the thermometer to take the temperature of some object of your choice--not your mouth! It may take some thought to come up with something else to measure the temperature of; be creative and check with other groups to see what object they used. Take several measurements and average them.

TEMPERATURE

reading 1	reading 2	reading 3	reading 4	AVERAGE
				T_4=

Use the equation of the actual line of best fit you found in TASK 5 to predict the voltage reading.

Voltage reading prediction: V_p = __________ mV

Now use the diode to measure temperature and record the voltage readings. Average them.

VOLTAGE				
reading 1	reading 2	reading 3	reading 4	AVERAGE
				V_4=

One indication of accuracy of this best fit straight line model is how well it predicts voltages.

The error in prediction is $|V_p - V_4|$ = ______________

VII. DISCUSSION

What contributes to the error calculated above? Discuss each contributing factor mentioned above including how significant you think it is in the calculated error.

The equation of the line reflects the type of relationship between temperature and voltage data. Examine the data and explain why the slope of the line is negative?

MATHEMATICS LABORATORY INVESTIGATION

TEMPERATURE AND VOLUME

Topic: **MEASUREMENT, GRAPHING, LINEAR REGRESSION**
Prerequisites: *Linear equations, calculator or spreadsheet graphing skills including finding intercepts.*

I. INTRODUCTION

Temperature has an effect on the size, hence volume, of many materials. Have you noticed that your fingers often swell in warm weather or that your rings may slip off in cold weather? Have you needed to let some air out of your tires during warm weather? Have you thought about why hot air balloons rise? Have you seen the effects of heat on rubber? Why did the engineers include expansion teeth in the surfaces of the overpasses around cities? What other effects of the change in temperatures have you noticed?

A hot air balloon rises because the air inside the balloon is a lighter density than the air outside the balloon envelope. Propane burners are used to heat the air inside. Increased temperature excites air molecules, speeding them up and causing the gaps between molecules to increase, resulting in the same air filling a larger volume. Filling a larger volume with the same amount of air molecules results in less density than the surrounding atmosphere and the balloon rises. Removing heat causes the molecules to slow down and return to their previous state and the balloon will sink. Balloon design engineers must understand the relationship between temperature and volume in order to properly size the propane burners required by a balloon system.

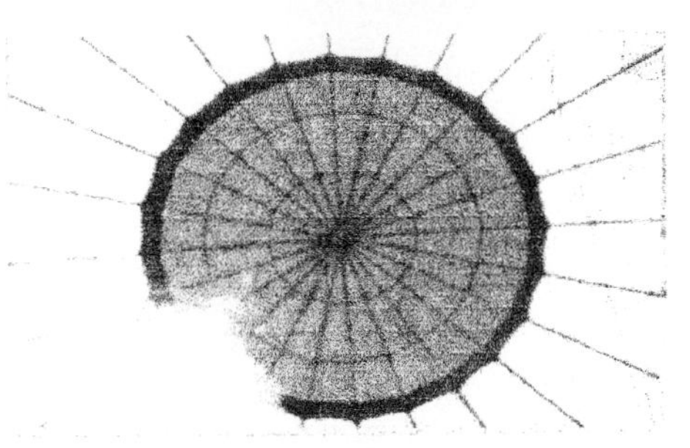

Meanwhile, air pressure exerted on the inside and outside of the balloon envelope fabric must remain constant; otherwise the balloon would not hold its shape. To allow constant pressure to be maintained and for the pilot to grossly adjust temperature, the balloon is slightly porous and also has designed openings.

In other applications, engineers use the temperature effect on gas, (i.e. cooling a gas causes the gas molecules to slow down) to filter minute impurities and to create superconductors with low electrical resistance. How slow can the gas molecules move? Theoretically the point where hydrogen molecules stop completely is called absolute zero. It is the zero reference point for the Kelvin temperature scale, and is very, very cold.

To understand why these things happen, you can explore the relationship between temperature and the dimensions of types of solids or gases. You will begin by measuring the volume of air in a plastic tube. Then you will measure the changes in the volume of the air as its temperature changes from very hot to cold (during the procedure the pressure must remain constant). After you have gathered and graphed the data , you will be asked to discuss the trends in the data and to describe this relationship. Finally you will be asked to write an equation that defines the relationship and to make predictions for other values of the temperature or the volume.

Equipment

4 or 5 feet of clear plastic tubing electrical tape
a bolt which fits snugly into the tubing
a tall container, a ring stand,
2 test tube clamps, 1 titration clamp
a metal dial thermometer (degrees Celsius)
a graduated metric cylinder , a meter stick,
a clear plastic straight edge
a source of hot water and ice, cooking oil
a bucket, a small funnel
looped wire tube manipulator, stirrer
graph paper
graphing calculator or graphing software

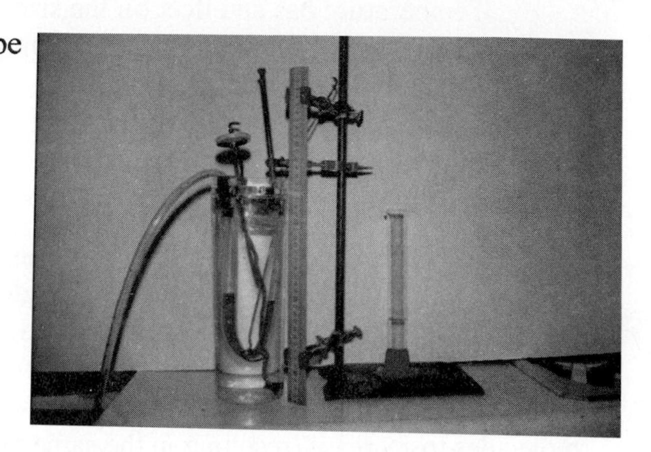

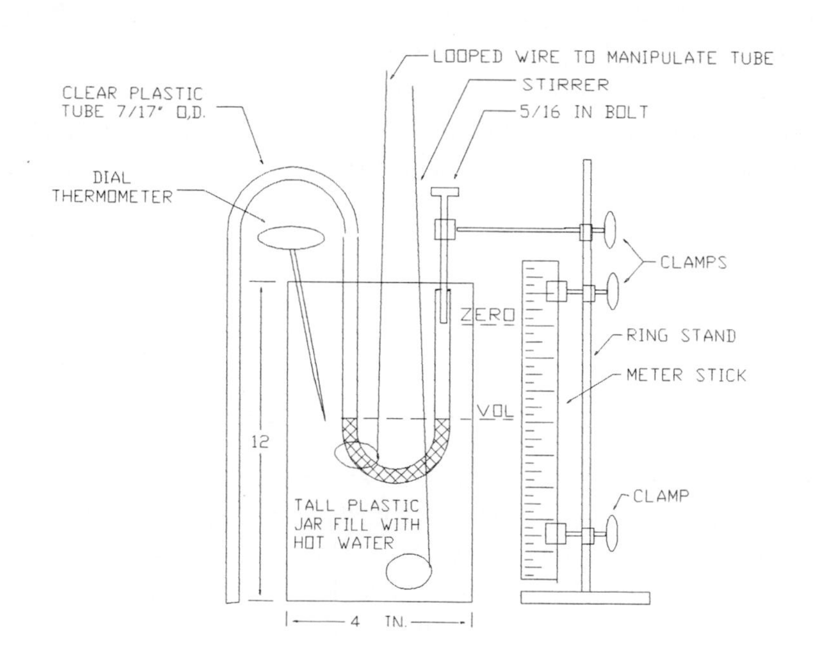

II. PREPARATION

Before beginning the data collection it is important to check out the measuring devices that will use.

What is the temperature of the room today?

What is the maximum error you can make using the meter stick attached to the apparatus?

Your instructor has calculated the volume per centimeter of tubing.

Record that volume here for use later on. V_t = ______ ml/cm

III. COLLECTING DATA

The lab has been designed to help you decide how the change in gas temperature affects the volume of a gas. A column of air has been trapped between the seal at the fixed end of the tubing and the cooking oil that has been placed into the free end of tubing. The lab requires that measurements be made of the temperature of the trapped air by measuring the temperature of the water surrounding the tubing. It also requires that measurements of the length of the column of trapped air be made and used with V_t to calculate the volume of the trapped air.

Task 1: Measuring and recording data.

Look at your apparatus and determine the length of the trapped air at room temperature. This experiment requires that pressure remain constant throughout the experiment. Pressure is constant when the oil levels in the fixed end and free end are equal. Use your looped wire and manipulate the free end of tubing so that the oil levels are even. Now you can measure the length of the trapped air at room temperature.

Record that length here. ________________

Add hot water to the beaker so that there is about an inch of water over the plugged end of the tubing. Put the thermometer into the beaker and wait a few minutes for the air in the tubing to adjust to the heat of the water. Thoroughly stir the water so that the temperature is uniform throughout the beaker. Record the temperature of the water. **Keeping the oil levels even in both ends of the tube**, measure and record the length of the captured air. As the temperature drops, the water will cool more slowly. Add ice to speed up the cooling process. You should calculate the volume of air while you are waiting for the ice to cool down the water.

Between measurements add 3 to 5 cubes of ice to cool the water down. Stir the water to keep the temperature uniform. When the temperature drops about 5 degrees, record the temperature and the new length of the captured air. Repeat this process at least ten times or as many times as possible.

	Temperature in degrees C.	Length of captured air in cm	Volume of air in ml
	T	L	V
1			
2			
3			
4			
5			
6			
7			
8			
9			
10			
11			
12			
13			
14			

What limiting feature of this model has made you stop collecting data?

Task 2: Filling in column 3 above.

Whether you use a scientific calculator, a graphing calculator with list functions, or a spreadsheet, you will need to calculate the volume of the air in the tubing based on your measurements for the length of the captured air in the tubing. The volume of air will be the distance between the end of the bolt and the top of the oil times the volume of the tubing per cm of length (V_t).

Using the symbols discussed thus far, write a formula below for the volume of the air.

Calculate the volumes and record them in your table.

IV. GRAPHING THE DATA

Next set up on graph paper a rectangular coordinate system for the Temperature-Volume relationship. Most often when graphing in this system, the values of the independent variable (the input values) are put along the horizontal axis. The values of the dependent variable (the output values) are put along the vertical axis. The dependent variable is said to be a function of the independent variable or the output, O, is a function of the input, I. It can be expressed as **f(input value) = output value** or **f(I) = O**. (Note that this is functional notation where the parenthesis do not imply multiplication.)

For this situation, **is the temperature or the volume to be considered the input value?**

____

Call the temperature T and call the volume V. **Use function notation to describe this situation.**

____

Since water boils at 100°C and freezes at 0°C, the values included for temperature are $0 \leq T \leq 100$. Look at your data and decide what values to include for volume. What is the minimum value possible for the volume of the air?

Can the temperature assume negative values? If not, why not?

Can the volume assume negative values? If not, why not?

Write an <u>inequality</u> statement for the values that you will use for your output variable.

____

When graphing, use appropriate labels as indicated in the diagram below.

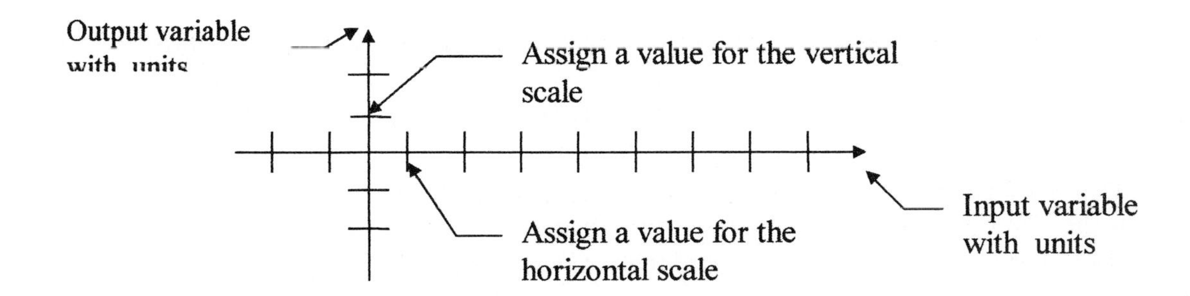

Task 3: Looking at trends in data from a scatter plot.

Plot the data points on graph paper.

In what order did you plot the points - (V, T) or (T, V)? ______

These data points are only a sampling of all the points that fit the situation. They form what is a *discrete* rather than a *continuous* collection of points.

Does this collection of data form any recognizable pattern? _____

If so, what pattern? ____

If you ignore a point or two, is the shape more easily recognizable? _____

Look at the trends in your data.

What seems to be happening to the volume as the temperature increases?

What seems to be happening to the volume as the temperature decreases?

Since both the temperature and the volume of the air really fluctuate continuously, the mathematical relationship is continuous and an unbroken line can be drawn to connect the data points.

Task 4: Roughly approximating a line of best fit.

Using your clear straight edge, can you find a line that most of the points come close to lying on? If so, lightly pencil it in. This line is an eyeball approximation of *the line of best fit*. Its importance is its ability to aid in predictions of other values.

Find the point on your approximation of the line of best fit where the temperature is 42 ° C and eye-ball over to your V axis to estimate the volume of air for that temperature. __________

Find the point on your approximation of the line of best fit where the volume is 4.8 ml and eye-ball over to your T axis to estimate the temperature of the air for that volume. __________

Estimate the values of the intercepts, that is estimate the volume when the temperature is 0° and estimate the temperature when the volume is 0 ml. Write your estimates as ordered pairs and include the appropriate units. (,) and (,)

This process can be done more precisely by writing an equation for the line of best fit and substituting values into the equation. The line of best fit can be arrived at using a variety of methods. A very rough estimate can be written by picking two points on your penciled in line and using algebraic procedures to write the equation of the line through those points.

Pick two points on the pencil line (**not data points**). **Write** an approximation of *the equation of the line of best fit* using algebraic methods.

____

Why is this only a rough estimate?

How do the T and V intercepts of your equation compare to the T and V intercepts of your graph?

Graphing calculators and software programs can quickly give you an equation for *the* line of best fit. *Linear regression* will yield an equation for the line of best fit.The process results in the lowest possible error by minimizing the sum of the squares of the vertical distances from the data points to the proposed line.

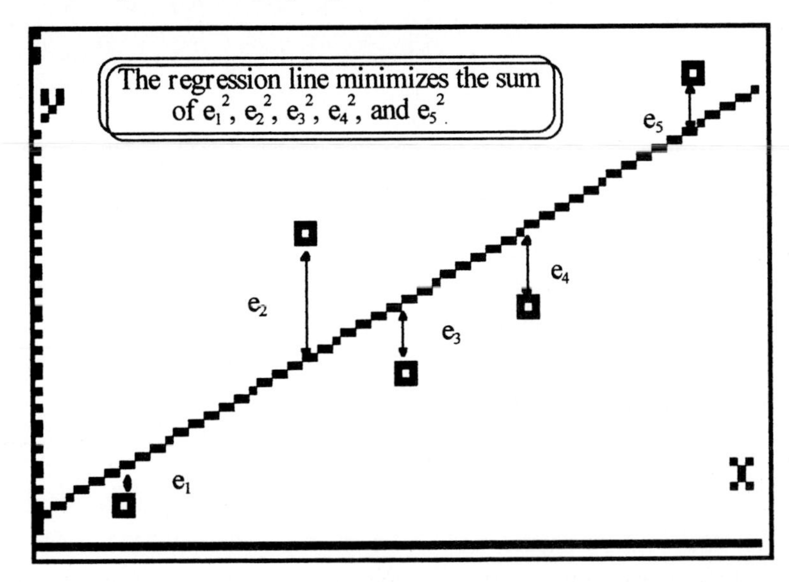
The regression line minimizes the sum of e_1^2, e_2^2, e_3^2, e_4^2, and e_5^2.
y
x
e_1
e_2
e_3
e_4
e_5

Task 5: Finding the line of best fit by calculator or computer software.

Write the equation in slope-intercept form for *the line of best fit.*

____

What is the slope of *the line of best fit*? ______________________

To what number of significant digits has the slope been calculated?

____

What units would be used for the slope? ___________________

How does it compare to the slope of your eyeball-drawn approximation?

How does it compare to the slope of the approximation you got algebraically?

Remember that another way to think of slope is as a rate of change. This rate of change is $\Delta V/\Delta T$.

Describe the rate in words.

What is the V-intercept? ________________

To what precision has the volume been calculated for a temperature of 0°C? ____________

Where should you round off approximations of the temperature? _________________

Where should you round off approximations of the volume?

____

Task 6: Using the line of best fit for predictions.

How might you use a graphing calculator and the line of best fit to fill in the table below?

Fill in the table.

Temperature in °C	Volume in ml
8	
74	
100	
30	
	1.105
	.84
	.53
	3.72

When predicting values from a line of best fit, in the geometrical sense, you can closely predict the value of any new point on the line. When using this process to predict other values for a given experiment, the only values that you can rely on to have a small percentage of error are the values that lie within the limits of the gathered data. **Predictions outside of the range of the original data are often made but further investigations sometimes prove them erroneous.**

What were the maximum and the minimum values of the temperature that you recorded?

____

What were the maximum and minimum values of the volumes recorded?

____

Go back to the table and check off the values which should have a relatively small percentage of error.

Which of the predictions can not be easily tested? Why?

Which of the predictions are unrealistic? Why?

V. - PREDICTING THE ABSOLUTE ZERO

Change the window on your calculator until you can see the T-intercept of the line of best fit.

What window are you using? ______________________________

Use the root function of your calculator or other software to find the value of T-intercept.

Describe, in terms of temperature and volume, the values of the T-intercept.

Give a physical interpretation to a volume of zero.

The value of the temperature when the volume of a gas is zero is called the *absolute zero.*

What value did you predict for "the absolute zero"? ____________

Write it as ± the number of degrees that you think you are off. ____________

What did other groups predict? ____________

What do you think contributed to the differences in the predictions?

MATHEMATICS LABORATORY INVESTIGATION

TOPICS IN STAIRCASE DESIGN

Topics: **ANGLES**
Prerequisite: *Trig Definitions, Spreadsheets*

I. INTRODUCTION:

Since ancient times, architects have been trying to design stairs with dimensions suitable to the human gait. Various rules of thumb have been employed but it was not until about 1762 that François Blondel, director of the Royal Academy of Architecture in Paris, first mathematically defined the rule of thumb that is still in use today. He observed that the normal human walking gait was about 24 inches and that this amount needed to be reduced by about two inches for every vertical inch of the stair.

Time has borne out his observation. When the Lincoln Center for the Performing Arts in New York City was built, the plaza staircase leading to the building had shallow risers (3 3/8") and deep treads (25"). This imposed such an awkward gait that numerous falling accidents occured, especially in descending the staircase. As a result, much of the staircase was replaced by a ramp system and the remaining stairs were closed off by railings.

When designing or making changes to the built environment, building codes are an essential part of the design process. the existence of building codes focuses on life safety: to ensure the proper exits, corridors, doorways to allow escape in the event of a fire or other hazard. Another major purpose is to provide the proper use of materials and finishes reducing the possibility of hazardous fumes and smoke.

There are various codes used throughout the country; these will vary from state to state. In Massachusetts, the Commonwealth of Massachusetts State Building Code, 5th Edition, 780 CMR (MSB) is the one currently in use, adopted by the State Board of Building Regulations and Standards. In addition to state and local building codes, there are regulations and laws that affect the design of spaces. These may include The Life Safety Code (published by the National Fire Protection Association), the Architectural Barriers Board and the Americans with Disabilities Act (ADA), which is not a code at all, but a comprehensive civil rights law governed by the Justice Department. The section of the ADA that pertains to designing the built environment is the Federal Register, Part III, 28 CFR.

Remember that codes and laws are written as minimum standards. It is up to the designer to create the best solutions from these requirements and not to rely solely on providing the minimum standard.

II. BUILDING CODES

Modern building codes are designed to provide a mathematical basis for appropriate stair construction. The following are example requirements for residential interior stairs:

Code A: (MSB 816.4.1) For residential interior stairs, the height of the risers cannot exceed eight and one quarter inches.

Code B: (MSB 816.4.2) All risers must be approximately the same height. Adjacent risers can differ by no more than 3/16" and the difference between the shortest and tallest riser can be no more than 3/8".

Code C: (MSB 816.4.1) For residential interior stairs, the depth of a stair tread must be at least nine inches.

Code D: (MSB 816.4.2) All treads must be approximately the same depth. Adjacent treads can differ by no more than 3/16" and the largest and smallest tread can differ by no more than 3/8".

Refer to the following picture for the definitions.

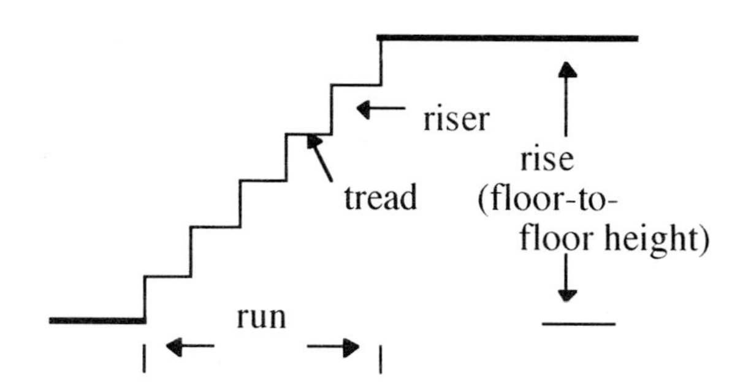

figure 1

Additionally, there are standards for comfort (which are **not** legal requirements or codes). One such standard, which, in this investigation, will be referred to as Blondel's Rule is:

Blondel's Rule: (*Architectural Graphics Standards*, by Ramsey and Sleeper) The height of a riser plus twice the depth of a stair tread should be between 24 and 25 inches.

III. STAIR CONSTRUCTION

TASK 1: Analysis of Staircase Runs

You are to construct a residential, interior staircase with a floor-to-floor height of 11 feet. Calculate the minimum number of risers necessary and the height of each riser.

Using Blondel's Rule, determine the range of allowable tread depths for the staircase described above. What would the minimum run (see figure 1) be for this staircase?

Create a spreadsheet with columns corresponding to the number of risers, riser height, minimum tread depth, and minimum run. The first row should correspond to the example computed above. On each subsequent row, the number of risers should be increased by one, stopping when the riser height becomes less than 4". Express all spreadsheet cells with an appropriate level of precision.

Hint: You will need to derive formulas that allow you to compute:

1. The riser height given the number of risers.
2. The minimum tread depth given the riser height.
3. The minimum run given the minimum tread depth.

On a graph, plot the minimum run versus the number of risers, using the number of risers as the horizontal axis.

What kind of a relationship does this graph seem to indicate? Explain your answer.

TASK 2: Angles

The slope of a staircase is defined to be the ratio of the floor-to-floor height to the run (rise/run). The angle of the staircase is the angle whose tangent is this slope.

Compute the slope and angle of the staircase you exhibited in TASK 1.

TASK 3: Staircase Angles

Extend the spreadsheet of TASK 1 by adding extra columns for the maximum tread depth and maximum run. The minimum and maximum runs will allow you to compute the maximum and minimum staircase angles, respectively. Add extra columns to compute these angles as well.

Graph allowable angles versus the number of risers, with the number of risers as the horizontal axis. (Above each value for the number of risers, there will be a vertical line from the minimum staircase angle to the maximum.)

For the staircases represented in your spreadsheet, what are all the possibilities for the number of risers of a staircase with an angle of 23 degrees ?

EXTENSIONS:

1. Consider a staircase with a handrail leading up to a landing where the handrail continues horizontally. In order for the handrail pieces to join properly, they need to be cut at the same angle. Describe this angle in terms of the angle of the stairs.

2. (To TASK 3): The stairs described in the spreadsheet are to be carpeted. Which of the possible staircases will use the fewest (most) linear feet of carpeting? In general, how do you determine the linear footage necessary to carpet a staircase?